THE
PLACE OF EMOTION
IN
ARGUMENT

THE
PLACE OF EMOTION
IN
ARGUMENT

DOUGLAS WALTON

THE PENNSYLVANIA STATE UNIVERSITY PRESS
UNIVERSITY PARK, PENNSYLVANIA

Library of Congress Cataloging-in-Publication Data

Walton, Douglas N.
  The place of emotion in argument / Douglas N. Walton.
        p.   cm.
  Includes bibliographical references and index.
  ISBN 0-271-00833-4 (alk. paper)
  1. Reasoning.   2. Fallacies (Logic)   3. Emotions.   I. Title.
BC177.W326   1992
168—dc20   91–30515
                                                           CIP

It is the policy of The Pennsylvania State University Press to use acid-free paper for the
first printing of all clothbound books. Publications on uncoated stock satisfy the
minimum requirements of American National Standard for Information Sciences—
Permanence of Paper for Printed Library Materials, ANSI Z39.48–1984.

For Karen, with love

CONTENTS

ACKNOWLEDGMENTS

The research that resulted in this book was supported by a Killam Research Fellowship from the Killam Program of the Canada Council, a Fellowship from the Netherlands Institute for Advanced Study in the Humanities and Social Sciences, a Research Grant from the Social Sciences and Humanities Research Council of Canada, and a Study Leave from the University of Winnipeg. The book was mainly written during a period when the author was a Fellow-in-Residence of the Netherlands Institute for Advanced Study in the Humanities and Social Sciences. My thanks go to the staff and fellows of NIAS in 1989–90 for their kindness in helping to make a foreign scholar feel at home in Holland and for providing a charming and stimulating ambiance for research work. As a member of the research group on "Fallacies as Violations of Rules for Argumentative Discourse," I would like to thank the other members of the group for discussions, comments, criticisms, and supportive advice: Frans van Eemeren, Rob Grootendorst,

Scott Jacobs, Sally Jackson, Agnes Verbiest, Agnes Haft-van Rees, Charles Willard, and John Woods.

For reading the manuscript and pointing out a number of errors, biases, and fallacies, I would like to thank Henry W. Johnstone, Jr., and Alan Brinton. Their comments helped me to give a better unity of development to the book and to catch some serious faults.

I would also like to thank Frans van Eemeren and Rob Grootendorst for permission to use parts of my paper "Types of Dialogue, Dialectical Shifts and Fallacies," forthcoming in the proceedings of the Second International Conference on Argumentation in the "Historical Background" section of Chapter 1.

Special thanks are due to Amy Merrett for information processing of the text and figures of the manuscript. Thanks also to Rita Campbell for doing the index, and to Harry Simpson for help with proofreading.

ADVICE TO THE READER

This book is built around the analysis and evaluation of case studies of controversial arguments in everyday conversations in Chapters 3–7. These case studies are inherently interesting examples of argumentation in their own right. They immediately provoke discussion and thoughtful reaction and for most readers will probably be the most exciting and rewarding part of the book. Their practical interest is immediately apparent, even to the reader who has no background or previous knowledge in the field of informal logic.

Chapters 1 and 2 present some tools useful for analysis and evaluation of these case studies and a background in the essentials of argumentation theory relevant to these cases. Chapter 8 welds the insights accumulated throughout the book into a theoretical synthesis, providing a theory of uses and abuses of the distinctive types of argumentation underlying the four traditional fallacies that are the chief target of analysis.

My advice, especially to the beginning (nonspecialist) reader, would be to skim over Chapters 1 and 2 first, for an understanding of the basic terms and concepts, and for the moment to just skip or glance over the parts that don't seem to be especially useful or interesting. Then start in reading seriously at Chapter 3, or any of the Chapters 3–7. Each of these chapters is relatively self-contained; you can actually read them in any order you find interesting. (The only exception is Chapter 7, which should probably be read after 3–6.)

Having gotten the real intellectual pleasure to be obtained from thinking over or discussing the case study materials of the middle five chapters, you can go back to ruminate over the more theoretical and methodological materials of Chapters 1, 2, and 8. The more you learn about the field of informal logic, the more these chapters will make sense and will clarify how this book fits into that field as a growing discipline with a historical background.

Another piece of good advice is not to read too much at one sitting. And if a part you have read seems unclear, stop there and read it over again the next day. The next morning it may suddenly become clear.

1

ARGUMENT AND FALLACY

The thesis of this book is that appeals to emotion have a legitimate, even important, place as arguments in persuasion dialogue, but that they need to be treated with caution because they can also be used fallaciously. The problem is that certain types of emotional appeals are very powerful as arguments in themselves, and they may have a much greater impact on an audience than is warranted in the case being argued. This book will study some particularly common and powerful arguments of this type in order to give critical guidance on how to confront and evaluate their uses in specific cases.

Two factors combine to enhance the trickiness of arguments that appeal to emotion. One is that an appeal to emotion may not be relevant, meaning that it may not contribute to the goals of dialogue that the participants in the argument are supposed to be engaged in. The other is that arguments based on emotional appeals tend to be weak arguments, based on presumptions rather than hard evidence. 'Weak' here means *logically* weak, that the premises do not support the conclusion strongly enough to fulfill the burden of proof. Such arguments become fallacious when the proponent exploits the impact of the appeal to disguise the weakness and/or irrelevance of an argument. As the Roman philosopher Seneca put it, appeals to emotion in argument cannot be trusted, because the emotions may be "moved by trifling things that lie outside the case" (*De Ira* I.XVII.5–XVIII.2). A weak or irrelevant argument can be taken as strong and relevant because of its powerful emotional impact on the respondent.

Four Fallacies

This book is about four types of arguments that have been traditionally treated as fallacies. What the four most obviously have in common is that they are all appeals to emotion. As powerful techniques of argumentation, all of them are based on a speaker's capability to rouse and exploit the sentiments and prejudices of a target audience. Another strong tradition is to contrast emotional thinking with "calm, dispassionate, logical reasoning," which could be called the idea of "cold logic" (Callahan 1988). This traditional separation of emotion and logic creates a climate of distrust toward any appeals to emotion in argument, making it easy to presume that the categorization of these four types of arguments as fallacies will be uncontestedly granted.

1. An *argumentum ad populum* is an argument that appeals to popular sentiment, or "to the people," to support its conclusion. It is also called "appeal to the gallery" or "mob appeal" in some textbooks. According to Engel (1976, 113), such "mob appeals" invite "people's unthinking acceptance of ideas which are presented in a strong, theatrical manner" and "appeal to our lowest instincts. . . ." According to Engel (114), such arguments are fallacious because they "steer us toward a conclusion by means of passion rather than reason."

2. An *argumentum ad misericordiam* is an argument that appeals to pity to support its conclusion. According to Michalos (1970, 51), this type of fallacy is committed "when one tries to persuade someone to accept a particular view by arousing his sympathy or compassion." He cites the

following example: "A student who missed practically every class and did nothing outside class to master the material told me that if he failed the course he would probably be drafted into the army." The reason this case is an *ad misericordiam* fallacy, Michalos commented (52), is that the issue is "not what happens if the student fails, but whether or not he deserves to fail." Appeals to compassion, Michalos concluded, are "stimulating, but irrelevant."

3. An *argumentum ad baculum* is an argument that appeals to a threat, or force, or fear, to support its conclusion (*baculum* means 'club' or 'stick' in Latin). According to Copi (1986, 99) the *argumentum ad baculum* is "the fallacy committed when one appeals to force or to the threat of force to cause acceptance of a conclusion." Copi cited Stalin's response when Churchill told the others at the Yalta meeting that the Pope had suggested that a particular course of action would be right: "And how many divisions did you say the Pope had available for combat duty?" Copi commented that *ad baculum* is "usually only resorted to when evidence or rational arguments fail."

4. An *argumentum ad hominem* is an argument that uses a personal attack against an opposing arguer to support the conclusion that the opposing argument is wrong. According to Fearnside and Holther (1959, 99), the *argumentum ad hominem* is a common, effective, and "odious" method of argument: "There is no argument easier to construct or harder to combat than character assassination, and this may be the reason personal attacks are so commonly on the lips of ignorance and demagogy." Fearnside and Holther do not think that the *argumentum ad hominem* is always fallacious, however, because personal considerations can be "relevant for judging the reliability of a man, his willingness to tell the truth." Hence the problem: how and when should one take the character of an arguer into account without committing the *ad hominem* fallacy? Character assassination is such a powerful tactic in argumentation that it is difficult to resist using it, difficult to defend against it effectively, and difficult to prevent the argument from degenerating into a personal quarrel once the tactic is used.

These four fallacies all rely on the prejudices in an audience; to stir them up a speaker directs an argument at what he or she takes to be the deeply held emotional commitments of the audience. Such tactics exploit the bias of an audience toward its own interests—whether it is financial interest, a social interest in belonging to a certain group (or a combination of these in an "interest group" or profession, for example), an interest in avoiding harm or danger, or a personal interest (as expressed in one's character or personal position as an advocate of a special cause).

Compared to the *ad hominem*, the literature on the *ad populum*, *ad misericordiam*, and *ad baculum* fallacies is scant. There are only a few articles, and what they say about these fallacies is tentative, sketchy, and incomplete, whereas more definite and systematic studies on the *ad hominem* fallacy have appeared, and they exhibit a promising degree of consistency, even if some basic questions of taxonomy and analysis have not yet been been answered. These studies of the *ad hominem* include Johnstone 1970, 1978; Barth and Martens 1977; Woods and Walton 1977; Flowers, McGuire, and Birnbaum 1982; Govier 1983, 1987; Brinton 1985, 1986, 1987; and Walton 1985, 1987a. The circumstantial type of *ad hominem* argument has been more thoroughly and comprehensively ana- lyzed by, for example, Johnstone (1970), Barth and Martens (1977), and Walton (1985). The largest gap is in the area of the so-called abusive or personal attack type of noncircumstantial, direct *ad hominem* argument; this appears to be less of a cognitively structured or propositional argument than the circumstantial type and more of an emotional fallacy, which raises fundamental issues about the personalization of argumentation.

In addition to one chapter each on the *ad populum*, *ad misericordiam*, and *ad baculum* fallacies, a chapter that concentrates on the personal and emotional aspects of the *ad hominem* fallacy has been included. Along the way, however, comments have been made on several of the other major informal fallacies where such commentary seemed relevant and, in some instances, necessary.

Two Other Arguments *Ad*

Two other types of arguments traditionally categorized under the heading of fallacies by the logic textbooks, in addition to the four types of arguments in the section above, make up the category of arguments *ad*. The word *ad*, meaning 'to,' 'against,' or 'toward' in Latin, seems to indicate that all six of these types of argument are directed "toward" something (like the emotions or frame of mind of the audience, or the person whom the argument was designed to convince, perhaps) other than evidence external to the participants in the argument.

The *argumentum ad ignorantiam* (argument to ignorance) is said by Copi and Cohen (1990, 93) to be "the mistake that is committed whenever it is argued that a proposition is true on the basis that it has not been proved false, or that it is false because it has not been proved true." This mistake is illustrated, according to Hamblin (1970, 43), by the argument that there must be ghosts because no one has ever proved that there aren't any. This seems intuitively to be a kind of error, just as if one were to try to prove a theorem in mathematics by arguing that lots of mathematicians had tried to prove it false but none had ever succeeded.

But Copi and Cohen are careful to note that not all arguments from ignorance are fallacious (1990, 94), citing the case where drug testing has demonstrated the absence of any toxic effect on rodents. Here the absence of any toxic effect is at least a practically useful, or presumptively acceptable, even if not conclusive, argument toward the conclusion that the drug is not toxic to humans.

The problem then is to determine when the argument is fallacious, or at least unreliable, and when it is not. But there is another problem as well. When the argument from ignorance is nonfallacious, what kind of "good" or "correct" argument is it? It does not look like any familiar kind of logical argument. If anything, it seems like a kind of epistemic argument that has to do with the knowledge base of a reasoner. Or perhaps it is a kind of practical argument, used to license a course of action as prudent—in the case above, on grounds of safety to human life—in a given situation, relative to what is known or can be reasonably assumed. But what kind of argument is that? Here the traditional logic textbooks have not been very helpful in giving guidelines for evaluation.

The *argumentum ad verecundiam*, which literally means the argument toward "shame" or "modesty," is usually taken to be an argument that rests on respect for authority. Judging from the examples often cited in the textbooks, it refers to the use of an appeal to expert opinion to support an argument.

Nowadays, with the advent of expert systems as a technology, it is the tendency to accept appeals to expert opinion in argument as reasonable. But as Hamblin notes (1970, 43), during some historical periods "arguments from authority have been especially disliked," and many traditional texts simply treated the argument from authority as a fallacy.

Copi and Cohen (1990, 95) begin by conceding that when trying to arrive at a decision, it is "entirely reasonable to be guided by the judgment of an acknowledged expert" who has studied a difficult or complicated matter thoroughly. But problems can arise, and the fallacy of *ad verecundiam* "arises when the appeal is made to parties having no legitimate claim to authority in the matter at hand." So the appeal to authority is, in principle, a legitimate kind of argumentation, but one that can go wrong, or be used wrongly, resulting in a fallacy.

This account, however, raises more problems than it resolves. An appeal to authority can go wrong in many different ways, even if the party appealed to does have a legitimate claim to being an authority in the matter at hand. So exactly when is such an appeal correct or fallacious in a given case of its use? And moreover, when such an argument is correct, what kind of "good" or "valid" argument is it? As an appeal to a third party, rather than directly to objective evidence, it seems a rather chancy and subjective kind of argumentation, depending on the person consulted. On the other hand, it is an extremely common kind of argumentation, widely used in all kinds of contexts of deliberation and discussion.

To throw some light on these basic questions about the interpretation of the six arguments *ad*, one should start with their historical origins.

Historical Background

Classification of fallacies is hazardous but has some value for initial guidance. The four emotional fallacies, the *ad populum, ad baculum, ad misericordiam,* and *ad hominem,* are easily grouped together because all obviously appeal to strong emotions in argumentation. But they are also a proper subset of the set Hamblin called the *ad* fallacies, which includes the *ad ignorantiam* and *ad verecundiam* as well as the other four.

Whately (1836; quoted in Hamblin 1970, 174) grouped all six of these arguments *ad* together. He saw them as arguments that are not necessarily fallacious but that are used fallaciously in some cases.

> There are certain kinds of argument recounted and named by Logical writers, which we should by no means universally call Fallacies; but which *when unfairly* used, and *so far as they are* fallacious, may very well be referred to the present head; such as the 'argumentum ad hominem,' ['or personal argument,'] 'argumentum ad verecundiam,' 'argumentum ad populum,' etc. all of them regarded as contradistinguished from 'argumentum ad rem,' or, according to

others (meaning probably the very same thing) '*ad judicium.*' These have all been described in the lax and popular language before alluded to, but not scientifically: the '*argumentum ad hominem,*' they say, "is addressed to the peculiar circumstances, character, avowed opinions, or past conduct of the individual, and therefore has a reference to him only, and does not bear directly and absolutely on the real question, as the '*argumentum ad rem*' does": in like manner, the '*argumentum ad verecundiam*' is described as an appeal to our reverence for some respected authority, some venerable institution, etc. and the '*argumentum ad populum,*' as an appeal to the prejudices, passions, etc. of the multitude; and so of the rest . . .

When Whately characterized all six of these *ad* fallacies as belonging to a special class of arguments, he contrasted them with the *argumentum ad rem* (argument toward the thing). He also added that the *argumentum ad judicium* was probably the same thing as the *ad rem*. What Whately apparently meant here was that all six of the *ad* fallacies listed have a 'personal' element, meaning that they are source-based in some way, directed at a source or person (a participant in argument) rather than at just "the thing" itself. They all have a "subjective" quality, as opposed to the "objective" evidence traditionally appealed to in argumentation. It could be said that all six of them have an underlying *ad hominem* aspect (see Johnstone 1970, who argued that all philosophical argumentation is *ad hominem* in nature).

In using the phrase *argumentum ad judicium*, Whately was referring to Locke's account. In a famous passage in his *Essay Concerning Human Understanding* (1690), in the chapter entitled "Of Wrong Assent, or Error," quoted in Hamblin 1970 (159–60), Locke listed four types of tactics for gaining assent. The *argumentum ad hominem, argumentum ad verecundiam,* and *argumentum ad ignorantiam* were collectively contrasted with a fourth type of argument: the *argumentum ad judicium*. As Hamblin notes (161), Locke indicated that he had invented three of these four terms himself. The term *argumentum ad hominem*, Hamblin conjectures, came from Aristotle's *Sophistical Refutations* (*De Sophisticis Elenchis*).

Before we quit this subject, it may be worth our while a little to reflect on *four sorts of arguments* that men, in their reasonings with others, do ordinarily make use of to prevail on their assent, or at least so to awe them as to silence their opposition.

First, The first is to allege the opinions of men whose parts, learning, eminency, power, or some other cause has gained a name and settled their reputation in the common esteem with some kind of authority. When men are established in any kind of dignity, it is thought a breach of modesty for others to derogate any way from it, and question the authority of men who are in possession of it. This is apt to be censured as carrying with it too much of pride, when a man does not readily yield to the determination of approved authors which is wont to be received with respect and submission by others; and it is looked upon as insolence for a man to set up and adhere to his own opinion against the current stream of antiquity, or to put it in the balance against that of some learned doctor or otherwise approved writer. Whoever backs his tenets with such authority thinks he ought thereby to carry the cause, and is ready to style it impudence in anyone who shall stand out against them. This I think may be called *argumentum ad verecundiam*.

Secondly, Another way that men ordinarily use to drive others and force them to submit their judgments and receive the opinion in debate is to require the adversary to admit what they allege as a proof, or to assign a better. And this I call *argumentum ad ignorantiam*.

Thirdly, A third way is to press a man with consequences drawn from his own principles or concessions. This is already known under the name of *argumentum ad hominem*.

Fourthly, The fourth is the using of proofs drawn from any of the foundations of knowledge or probability. This I call *argumentum ad judicium*. This alone of all the four brings true instruction with it and advances us in our way to knowledge. For: (1) It argues not another man's opinion to be right because I, out of respect or any other consideration but that of conviction, will not contradict him. (2) It proves not another man to be in the right way, nor that I ought to take the same with him, because I know not a better. (3) Nor does it follow that another man is in the right way because he has shown me that I am in the wrong. I may be modest and therefore not oppose another man's persuasion; I may be ignorant and not be able to produce a better; I may be in an error and another may show me that I am so. This may dispose me, perhaps, for the reception of truth but helps me not to it; that must come from proofs and arguments and light arising from the nature of things themselves, and not from my shamefacedness, ignorance, or error.

It appears then that origin of the much-used terms *ad ignorantiam*, *ad hominem*, and *ad verecundiam* in the modern logic textbooks derives from this passage in Locke. However, Locke's description of them as assent-producing devices used in an interactive framework where one person is "reasoning with" another seems quite different from the way they have been traditionally described in the logic textbooks, as fallacies. Locke's way of contrasting the first three with the *argumentum ad judicium* seems to characterize these three as a class. It will be shown in this book that his remarks are important to understanding all six of the arguments *ad* as distinctive subspecies or special techniques of practical argumentation. All six are appeals to something other than objective or reproducible evidence (*argumentum ad judicium*).

The word 'appeal' is important in characterizing arguments *ad*, as Sidgwick (1914, 145) accurately noted in summing up these six types of argument.

> *Argumentum ad* so and so; e.g. ad *baculum*, ad *hominem*, ad *ignorantiam*, ad *misericordiam*, ad *populum*, ad *verecundiam*. "Argumentum" may here be translated "an appeal"; e.g. an appeal to force, to a man's own professions or admissions, to ignorance, to pity, to popular views, to respect for authority. The second, third, and sixth in the above list were contrasted by Locke with the *argumentum ad judicium*, which he described as "the using of proofs drawn from any of the foundations of knowledge or probability. . . . This alone . . . brings true instruction with it and advances us in our way to knowledge."

All six types of arguments are "appeals" directed *toward* another party in a dialogue. Rather than appealing to objective evidence that is true or false independent of the given commitments of other participants in the dialogue, all six of these arguments appeal to something in the 'mind' of another party in the dispute—more specifically they appeal to the state of knowledge or the set of commitments of another party. Hence, they are all source-based or inherently subjective as types of arguments.

As Hamblin's commentary (1970, 160–63) shows, Locke did not say that these three arguments *ad* he cited were always fallacious. He seemed to leave it open, indicating that they could be used fallaciously as well as nonfallaciously. But he didn't pin down what was special about them as arguments *ad*.

Whately went further. Whately's program was to analyze all six of the arguments *ad* as fallacies that create presumptions and shift burden of proof in argument. Hamblin commented (175) that this "excellent and interesting" analysis "needs to be worked out in more detail," but he added that so far, this has not been realized and that none of the subsequent logicians (Schopenhauer, Mill, DeMorgan, or Sidgwick) "added anything very new to the study."

All of the arguments *ad* are basically tactics or mechanisms used to gain assent by an appeal to a party in the dialogue rather than an appeal to external evidence (*ad judicium*). The object of the argument is to shift a burden of proof, or a weight of presumption, toward the other party's side and away from the arguer's side. The way this is done may appear somewhat mysterious and unfamiliar to the modern reader schooled in deductive logic; it is quite different from the more familiar methods of amassing data or objective evidence "arising from the nature of things themselves." Intriguingly, Hamblin suggests (161) that the *argumentum ad judicium* is "surely close" to Aristotle's concept of a demonstrative argument, whereas the *argumentum ad hominem* is reminiscent of Aristotle's concept of a dialectical argument.

These remarks are provocative but, by themselves, not very helpful in pinpointing the real nature of arguments *ad*. Locke clearly meant to indicate that they involve dialectical, or interactive, argument exchanges, in which two parties "reason together." Whately pushed this provocative idea further by saying that they employ the passing back and forth of a burden of proof or weight of presumption in such an exchange. But why, how, and under what conditions are such arguments used "correctly" in a dialogue exchange? And how can one tell when such an argument has "passed over the logical line" and become fallacious? These very puzzling and provocative questions remain open.

The underlying problem with all six arguments *ad* is that they involve a different kind of reasoning, or use of reasoning, in argument than customarily found in logic.

Three Concepts of Reasoning

Until the mid-twentieth century, the central concern of logic was with the theory of deduction, the inference mechanism whereby one proposition (the conclusion) is inferred or deduced from a set of propositions (the premises). According to the theory of deductive logic, these premise propositions are "assumptions" or "axioms." They can be freely accepted or rejected, and in principle, no restraints or conditions limit which propositions can play this role in the theory. This theory takes a semantic approach in focusing on the truth or falsity of the propositions in an argument.

With the advent of AI, or "artificial intelligence," the thinking and reasoning of machines in computer science, the central concern of the study of reasoning has changed to knowledge-based reasoning (Wilensky 1983). This approach is concerned not only with the inference mechanism but also with where the premises of an inference come from. In knowledge-based reasoning, the premises are drawn from a specific domain: the knowledge base of the system or reasoner that is drawing a conclusion. The operations of deducing (inferring) conclusions are carried on within the reasoning system itself, which contains the knowledge base of the system. Here the set of propositions playing the role of premises cannot be any arbitrarily selected propositions or "assumptions." The system is now bounded to a special set of premises—not just any arbitrarily designated proposition can be a premise.

The idea of knowledge-based reasoning leads to a third concept of reasoning, the next step in the evolution of logic: that of *interactive reasoning*, where two separate systems (participants) interact in reasoning and where each participant draws premises from the knowledge base of the other and infers conclusions from them.

The first approach, the theory of deduction, is a linear system that draws a single conclusion from an arbitrary set of premises. Within this framework, there may be alternative, equally acceptable paths. Which among them is the actual path used to get from the set of premises to the conclusion may not be so significant. If two proofs exist, the shorter one is usually preferred, but only on aesthetic grounds. The important thing is whether a solution exists or not.

The second approach, that of knowledge-based reasoning, begins with a bounded set of premises called the knowledge base of the system, to which

DEDUCTIVE REASONING

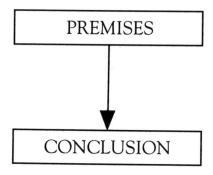

Fig. 1.1

a question is put by a user: "Is this proposition (the conclusion) known to be true?" The system then "chains backward" to see whether (and how) this conclusion can be inferred from the premises contained in its knowledge base. One problem here is that the chain of deductive steps may be long, and several solutions may exist, so the system must be concerned with choosing the best route for deducing the conclusion from the premises (Waterman 1986). How to deduce the conclusion from the premises becomes an important factor in searching through a knowledge base for a conclusion queried.

By this second approach, different routes of inference from the premises to the conclusion must be compared and evaluated. And it also becomes important in some cases to "reason backward" from the conclusion to the premises (Walton 1990). The approach of knowledge-based reasoning, therefore, is not so linear as the pure theory of deduction. How one gets from one point to another becomes critical.

Here it becomes important to map out the different avenues the system can use in following inferences backward and forward, to and from a conclusion. How the system searches through the internal stock of propositions it contains (its knowledge base) in order to derive a conclusion is the "reasoning" of the system. This second concept of reasoning is more complex than the first one, because "where you come from" (the

KNOWLEDGE-BASED REASONING

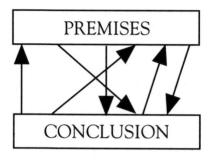

Fig. 1.2

premises) and "how you got there" (the heuristics, or search paths) are important.

In the second concept of reasoning, however, the system (the reasoner) is self-contained, with two exceptions. In dynamic systems, new knowledge can come into the system or go out of the knowledge base in the system. Also, the system is often viewed as having a user interface, so that an external participant can ask questions or convey facts to the system. This concept of a user interface leads to a third approach to reasoning.

According to the third concept, called the *interactive* (*dialectical*) concept of reasoning, two separate systems (participants in reasoning), each with its own knowledge base, reason together, one system interacting with the other. Each system takes account of the knowledge base of the other, as the two systems interact in joint reasoning. The interaction can take the form of transfering propositions from the one knowledge base to the other, or it can take the form of questioning, where one system advances a proposition to the other in the form of a query, for example, "Do you accept this proposition as true?" or "Can you prove that this proposition is true?" What makes dialectical reasoning distinct from the first two kinds is that the conclusions inferred by a system follow from premises drawn from the propositions in the knowledge base of another system.

Accordingly, interactive, or dialectical, reasoning characteristically takes the form of a dialogue—a sequence of question-reply interactions between two participants (systems, knowledge bases). Let us call two such participants in dialogue White and Black. White reasons by inferring a conclusion from a set of premises made up of propositions accepted by Black. And Black reasons by inferring a conclusion from a set of premises made up of propositions accepted by White. How each one reasons depends, therefore, on what the other one knows or accepts as true (Walton 1984).

INTERACTIVE REASONING

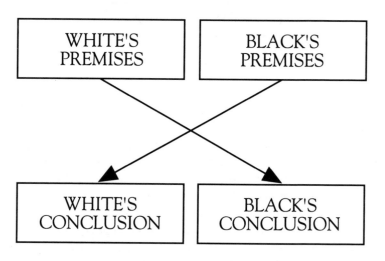

Fig. 1.3

The distinguishing characteristic of interactive reasoning is that the reasoning is directed toward another party. By contrast, the deductive and knowledge-based concept of reasoning require only a single, self-contained system, where the sequence of reasoning is essentially internal to the system itself. The first two concepts of reasoning are monolectical, in the sense that only a single reasoner or system is involved. By contrast, interactive reasoning is inherently dialectical, in that it involves a back-and-forth sequence of interactions—in effect, a dialogue between two participants who are reasoning together.

Different forms of interaction can take place between two parties engaged in the dialogue of interactive reasoning. For example, information-seeking dialogue, where the first party has the goal of finding out something the second party knows, involves a different kind of interactive reasoning from persuasion dialogue, where the first party has the goal of showing the second that the point of view of the first is right.

With the advent of AI, knowledge-based reasoning has begun to supplant the narrower theory of deductive reasoning that had, from the Renaissance to the present, been accepted as the dominant paradigm of reasoning. As knowledge-based reasoning becomes more widely accepted as a respectable concept, the way is pointed toward the third concept of interactive reasoning. Interactive reasoning lies in the future. But curiously enough, it also lies in the past. Although it is alien today, the idea of two-person interactive reasoning in dialogue was widely accepted by the ancient Greeks.

Perhaps they were aware that the interactive model is useful for solving problems in the critical evaluation of argumentative discourse. For example, it is a common problem to confront an example of argumentation, looking for evidence in the text of the dialogue to support or undermine the selection of propositions as the premises or conclusion of a disputed argument. To merely define premises or conclusion as "given" or "designated" propositions would leave the problem unresolved and unreasonable. The semantic concept, by itself, is too narrow to address these kinds of problems.

Where the semantic approach to argument evaluation has especially fallen short is in the area of informal fallacies, characteristic deceits or tricks of persuasion in faulty arguments. It has been extensively documented, in Hamblin 1970 and Walton 1987b particularly, how the analysis of the fallacies requires a pragmatic dialogue structure of reasoned argument that goes beyond the narrowly semantic concept of logic. But Hamblin's chapter on argument (1970, chapter 7) also indicates that the required pragmatic structure of dialogue carries with it the necessity of rethinking the traditional categories of argument evaluation in logic, if difficulties and puzzling questions related to the concept of argument are to be clarified and straightened out.

The Concept of Fallacy

In the standard textbook treatment—see Hamblin 1970 (chapter 1)—a fallacy has come to be understood as an argument that seems to be valid but is not. In this understanding of the concept of a fallacy, 'seems' refers to the psychology of the audience, or respondent, to whom the argument is directed. And the term 'valid' suggests a semantic evaluation, oriented towards deductive validity of an argument as the principal property in question. This psychological-semantic concept of fallacy has not been a very successful framework for understanding the fallacies or dealing with them in a practical and constructive way (Walton 1987b).

The present monograph takes a pragma-dialectical view of fallacies as misused techniques of argumentation that go against the legitimate goals of the dialogue the participants in a given argument are supposed to be engaged in. This general approach has been advocated by van Eemeren and Grootendorst (1984), who see fallacies as violations of the rules of a critical discussion. It is also advocated in Walton 1984, where fallacies are viewed as incorrect moves in a persuasion dialogue where the goal of one participant is to persuade the other that a designated thesis is a commitment the other should accept.

According to the pragma-dialectical approach (the term 'pragmatic' means 'practical'), an argument is seen as a dynamic exchange, a sequence of pairs of speech acts carried out by two participants engaged in dialogue. A speech act is a verbal means of performing a particular function in a dialogue, as described by the felicity conditions required for an appropriate performance of the act. For example, the speech act of requesting constitutes an attempt to get a hearer to perform an action (essential condition); as such it predicates the future action of an addressee (propositional content condition) and specifies that the speaker wants the addressee to carry out the requested action. An added condition (the preparatory condition) specifies that the speaker believes the hearer is able to carry out the requested act, that the speaker has reason to want the act performed, and that the hearer would not (obviously) do the act anyway.

The pragma-dialectical approach is pragmatic, or practical, because a speech act is carried out for a purpose in the context of dialogue. It is dialectical, or interactive, because performing a speech act requires two parties, a speaker and a hearer, who are engaged in a dialogue.

This dialectical approach to the study of fallacies was exemplified in

Aristotle's treatment of sophistical refutations in his *Topics, Rhetoric,* and *De Sophisticis Elenchis.* For Aristotle, refutation was a kind of argument move used by one party to defeat another party when the two parties were engaged in reasoning together in a sequence of interactive questions and replies. A sophistical refutation for Aristotle was an intentional tactic or trick of argumentation that could be used to deceptively and unfairly refute an opponent in dialogue (Kapp 1942; Evans 1977).

The English word "fallacy" is based on the Latin *fallacia,* which was in turn the offshoot of Aristotle's concept of a sophistical refutation. But many linguistic unclarities and uncertainties have compounded the confusions engendered by the evolution of the word "fallacy" into its modern meaning. According to Hamblin (1970, 50), Greek had no precise synonym for "fallacy," and the related English terms "sophism" and "paralogism" are often used as equivalents in the translation of Greek terms used by Aristotle.

The Latin word *fallacia* means "deceit, trick, artifice, stratagem, craft or intrigue" according to Lewis and Short (1969, 721). *Fallacia* is descended from the Greek verb *sphal,* meaning "to cause to fall," either in sports competition or by verbal tactics of argumentation. Interestingly, this meaning is quite consistent with Aristotle's idea of a sophistical refutation as an intentional deception or trick of argument used to defeat a partner in dialectic by "causing him to fall," that is, to be refuted in argument. But even though, as Hamblin (1970) showed, the fallacies evolved from Aristotle's list of sophistical refutations into the classifications of fallacies in the modern logic textbooks, the pragma-dialectical roots of the concept in Greek philosophy faded into obscurity in modern times.

However, the concept of fallacy that is the basis of the present book is somewhat different from Aristotle's concept of a sophistical refutation. Aristotle's notion of a sophistical refutation as a deceptive tactic of argument used intentionally by one participant in a dialogue to unfairly get the best of the other is a much better concept than the modern idea of fallacy as an argument that "seems valid but is not," because it is a fully dialectical concept. But it is somewhat too radical in one particular respect: it requires a fallacy to be an intentional deception.

It is important that intention to deceive not be a requirement for the commission of a fallacy. For this requirement would tie the fallacy in a given case too closely to the real psychological intent of the alleged perpetrator. Instead, it is better to tie the existence of a fallacy to a normative reconstruction of the arguer's commitments, based on (a) the goals and rules of the dialogue the participants are supposed to be engaged in and (b) the interpretive reconstruction of the given text of discourse in a particular case. In other words, the relevant evidence should be derived from what the participants actually said, in light of the verbal exchange they were supposed to be engaged in.

According to this new pragma-dialectical concept, a fallacy is a technique of argumentation that may in principle be reasonable but that has been misused in a given case in such a way that it goes strongly against or hinders the goals of dialogue. The major fallacies involve the use of argumentation schemes that can be used correctly in dialogue, but when they are used correctly, they are weak, presumption-based arguments, used to shift a burden of proof. These arguments can also be used incorrectly, in violation of the rules of dialogue, to commit fallacies. The problem then is one of distinguishing between the fallacious and nonfallacious instances of use.

A burden of proof is a "weight," or degree of strength an argument needs to accumulate in order to prove its conclusion in a dialogue. Normally if you assert a proposition in a critical discussion, you are obliged to prove it, if challenged, meaning that you take on the burden of proof appropriate to that proposition in the given context of dialogue. But in argument, presumption is a different kind of speech act from assertion.

In the speech act of presumption, the burden of proof shifts to the side of hearer as a burden to refute the proposition in question; it no longer rests on the side of the proponent, who has put that proposition forward in the dialogue (Walton 1988). Presumptive reasoning is *defeasible*, meaning that it is subject to defeat should new evidence come in during the course of a dialogue or be introduced by any participant. Defeasible reasoning is inherently tentative—it holds only temporarily, always subject to the possibility of rebuttal.

To evaluate fallacies is to judge how the technique of argumentation has been used in a particular case. One needs, first of all, to understand generally how the technique is used correctly and how it can be used incorrectly to commit a fallacy. These tasks require an understanding of the normative models of dialogue and their application to a particular case. Each case must be judged on its merits, in relation to the textual and contextual evidence that is available. If not enough evidence is given to "pin down" the charge that a fallacy has been committed, the claim of fallacy must be evaluated in a conditional manner. Such conditional judgments are often interesting and informative, even if they fail to pin down a fallacy.

A fundamental thesis of this book is that the appeals to emotion studied here are inherently weak arguments based on presumptive reasoning. Presumptive reasoning is always defeasible in that it shifts a weight or burden of proof in dialogue but is inherently open to rebuttals. From this it does not follow, however, that such arguments are always fallacious. Yet these presumption-based arguments can sometimes be taken as much stronger or more decisive in settling an issue than they properly should be. When this occurs, fallacies can arise, especially when the proponent of the argument tries to exploit the apparent decisiveness created by their emotional impact.

Another thesis coming out of the book is that when these emotional appeals are reasonable arguments, they are often instances of practical reasoning, directed toward a conclusion describing a prudent course of action. *Practical reasoning* may be defined as a kind of goal-directed, knowledge-based, action-guiding argumentation that steers an agent toward a prudent course of action in a set of particular circumstances (Walton 1990). Of course, in practical reasoning, particular circumstances tend to be variable rather than fixed, and therefore practical reasoning tends to be presumptive reasoning as well.

The deductive-semantic point of view, which has been so prevalent in logic since Aristotle, tends to see appeals to emotion as fallacious, because this practical aspect of reasoning is overlooked or ignored. It is hoped that the pragma-dialectical approach of this book will redress this imbalance and to some degree vindicate appeals to emotion in argumentation as having practical value.

Types of Dialogue

A *dialogue* is an exchange of speech acts between two speech partners in turn-taking sequence aimed at a collective goal. The dialogue is *coherent* to the extent that the individual speech acts fit together to contribute to this goal. As well, each participant has an individual goal in the dialogue, and both participants have an obligation in the dialogue, defined by the nature of their collective and individual goals. A *discussion* is a particular type of dialogue that has the goal of arriving at the truth of a matter or clearing up difficulties related to a problem.

In some dialogues the goal is to prove something, and in this type of dialogue a primary obligation is the burden of proof. A burden of proof is a weight of presumption allocated ideally at the opening stage of the dialogue, set for practical purposes to facilitate the successful performance of the obligations of the participants during the course of the dialogue (Walton 1988). The device of burden of proof is useful because it enables discussion to come to an end in a reasonable time.

One important type of dialogue is the critical discussion, well described by van Eemeren and Grootendorst (1984), which is a type of *persuasion dialogue*, in which the goal of each party is to persuade the other party to accept some designated proposition, using as premises only propositions that the other party has accepted as commitments. The goal of the critical discussion is to resolve a conflict of opinions by means of rational argumentation. The concept of commitment is the basic idea behind all dialogue as a form of reasoned argumentation—see Hamblin 1970, 1971; Walton 1984; and Walton and Krabbe 1990.

In *information-seeking dialogue* the goal is for information to be transmitted from one party to the other. The *interview* is one type of information-seeking dialogue. In Hamblin's formal game of dialogue (H)—see Hamblin 1970 (256–71)—transfer of information is ostensibly the goal, but there seem to be implicit aspects of the persuasion dialogue involved as well (even though Hamblin did not formulate explicit win-loss rules). Another type of information-seeking dialogue is the *advice-solicitation dialogue*, where the goal of the one party is to seek advice, in order to carry out an action or solve a problem, by consulting another party who is in a special position to offer such advice. This type of dialogue is less adversarial in nature than persuasion dialogue.

An important type of advice-solicitation dialogue is the *expert consultation dialogue*, where a nonexpert in a domain of skill or specialized knowledge consults an expert in order to get the expert's opinion or advice in a form he can use to solve a problem or go ahead with a course of action in an informed and intelligent way (see De George 1985; Waterman 1986). In this type of dialogue the expert respondent has an obligation to offer her best advice in clear and accessible language, at the same time admitting her limitations and doubts. The obligation of the advice-seeker is to ask clear and specific questions that the expert can answer in relation to the problem at hand. A commonplace case is physician-patient dialogue in medical treatments and consultations.

In *negotiation dialogue* the aim for both parties bargaining over some goods or interests is to "make a deal" by conceding some things while insisting on other things. Each side tries to figure out what the other side wants most, or feels is most important, of the goods at stake. Negotiations, like union-management bargaining for example, are now often guided by professional mediators. Fisher and Ury (1983) describe methods of negotiation and mediation studied in the Harvard Project. Donohue (1981b) presents a theory of negotiation as normative structure of rule use in interaction. Donohue (1981a) gives an analysis of the kinds of tactics used by participants in negotiation dialogue.

In *inquiry dialogue* the goal is for the participants to collectively prove some particular proposition, according to a given standard of proof, or to show that the proposition cannot be proved at the present state of knowledge. The intent of the inquiry is to be *cumulative*, to work only from established premises that will not require further discussion or retraction once set in place at the appropriate stage of the inquiry. The sequence of the inquiry is meant (ideally) to be linear or branching in one direction, so that circular argumentation can always be excluded.

The *scientific inquiry*, called the *demonstration* by Aristotle, requires that proof proceed only from premises that are either axiomatic or that can be established by methods of inference accepted by standards in a particular branch of scientific knowledge. Other inquiries, like public inquiries into air disasters and the like, rely on expert testimony of scientific consultants. But the goal is the same. The purpose is to prove a conclusion by a high standard of proof that eliminates unverified presumptions from the line of advance.

The *quarrel* is a type of dialogue where the goal of each participant is to "hit out" verbally at the other, and if possible, defeat and humiliate the other party. The quarrel is typically precipitated by a trivial incident that "sparks" an escalation of emotions, with both parties adopting a stubborn or "childish" attitude during the argumentation stage. The real purpose of the quarrel is a cathartic release of deeply held emotions so that previously unarticulated feelings can be brought to the surface—feelings that would not be appropriate to bring out for discussion in the course of a normal, polite, public conversation.

In contrast to the critical discussion, neither participant in the quarrel is really open to changing his position, even when confronted by convincing evidence and reasonable arguments. The quarrel is not a good friend of logical reasoning; it is characterized by caviling, by a brushing aside of logic, and by giving priority to the need to defeat the other party crushingly, by any means, foul or fair, that comes to hand. The quarrel is typified by excesses.

The quarrel is a species of *eristic dialogue*, a type of dialogue that is almost purely adversarial, where finding the truth of a matter and paying attention to logical reasoning procedures are always subservient to winning out over the other party (Kerferd 1981, 59). Aristotle (*De Sophisticis Elenchis* 171 b 25) compared contentious (eristic) argument to unfair fighting in an athletic contest where the participants are bent on winning at any cost and stop at nothing. Those who behave like this merely to win a victory are "contentious and quarrelsome." Aristotle related the quarrel to sophistry in an interesting way as well (171 b 27): those who are contentious merely to win a victory are "quarrelsome," but those who do so for the sake of reputation for wisdom and the consequent payment are "sophistical" (171 b 25-35). The art of the sophist, wrote Aristotle, is to achieve the goal of apparent wisdom. Thus, according to Aristotle (171 b 33), quarrelsome people and sophists use the same kind of arguments, but for different reasons.

It is often presumed that the quarrel is a wholly bad type of dialogue, best avoided altogether. But this point of view overlooks some valuable benefits of the quarrel. By allowing powerful feelings to be expressed through the articulation of deeply held grievances, the quarrel can improve mutual understanding and cement the bonds of a personal relationship. A quarrel can split two people apart, but if it has a good cathartic effect, it can function as a substitute for physical fighting and draw people closer together in the course of a meaningful relationship.

The critical discussion, on the other hand, enables a participant to articulate and clarify deeply held commitments on an issue by testing arguments in adversarial dialogue with an able critic. This maieutic function brings dark commitments to light by exposing them to critical questioning.[1] The cathartic function of the quarrel generates more heat

---

1. The term "maieutic" refers to the bringing to light of new ideas by one party who engages in dialogue with another party. The source of the term is the Greek word *maieutikos*, an adjective meaning "skilled in midwifery." In the *Theatetus* Plato portrayed Socrates as performing the function of midwifery by asking critical questions in dialogue with another party, assisting that party to bring forth new ideas in the discussion.

than light, but the resulting expression of repressed feelings can still be a valuable side-benefit of this type of argumentation.

Dialectical Shifts

Emotional appeals and the personalization of arguments are not necessarily impediments or defects in a critical discussion. Indeed, the personalization of one's argument, by addressing it to the personal (dark-side) commitments of the other party, is a necessary requirement for the fulfillment of the maieutic function of a critical discussion. Thus, in some cases, at any rate, it would seem that emotional appeals like the *ad hominem* or *ad populum* arguments are nonfallacious.

Fallacies often arise where there has been a dialectical shift, a change from one context of dialogue to another. For example, if a dialogue originally started out as a scientific inquiry but then shifted to a critical discussion, this shift would very much affect how an *ad hominem* argument that was advanced during the sequence of dialogue should be evaluated. If the inquiry was properly closed off and the *ad hominem* argument took place entirely in the critical discussion part, then the argument could be nonfallacious. Whereas if the shift was an illicit one, not properly agreed to by both parties, then the *ad hominem* argument should be evaluated by the rules and requirements of the original dialogue that the participants were supposed to be engaged in—the inquiry. By these standards, the *ad hominem* argument could definitely be fallacious.

Shifts also explain "seeming validity" of fallacies. An argument that would be seen as outrageously unconvincing in an inquiry, for example, could appear much more plausible (and be much more deceptive) if there has been a subtle shift to persuasion dialogue.

The essential difference between the critical discussion and the inquiry is reflected in both the goals and methods of these two types of dialogue. The goal of the inquiry is to prove some particular proposition. Therefore the method of the inquiry is to gather all the relevant evidence, drawing a conclusion from premises that can be firmly established. The goal of the critical discussion is to resolve a conflict of opinions by looking at the arguments on both sides. The method is to raise critical questions to explore an issue so that the strongest arguments on both sides can be examined. One proceeds by question and reply. The other proceeds by a buildup of well-established evidence, with the aim of eliminating the need for further questioning or retraction of premises.

Aristotle pinpointed this essential difference by distinguishing between the methods of the two types of argumentation frameworks he called demonstration and dialectic. According to Aristotle (*De Sophisticis Elenchis* 172 a 20), dialectic proceeds by interrogation, by questioning premises. Demonstration proceeds by proof, from premises that are well established. The assumption in this distinction is that the aims of demonstration and dialectic are essentially different, and that therefore their methods must reflect this difference. Any dialogue where the goal is to prove something, in the sense of pinning the conclusion firmly down, must refrain from continual questioning. In other words, the basic premises and principles (axioms) cannot be subject to continual questioning (within the demonstration itself, at any rate). The need for continual retraction as the investigation proceeds would interfere with the cumulative progress of proving the conclusion.

Another difference cited by Aristotle (172 a 30) is that the art of dialectical examination requires no knowledge of any definite subject or field of expertise. Aristotle writes (172 a 15), "Dialectical argument has no definite sphere, nor does it demonstrate anything in particular." This remark indicates another distinction between the inquiry and the critical discussion. The inquiry can, in many cases and even typically, be bound to subject-matter specialization. It may be a scientific inquiry where only technical experts can take part. Of course, it is also possible to have "public" inquiries, for example, into the cause of an air disaster. But even there specialists have roles to play in parts of the inquiry. Nevertheless, everybody engages in critical discussions, whether they are scientific experts or not.

A critical discussion can have input from experts to provide advice or information, but the critical discussion is not bound to subject-matter specialization or expertise in the same way that the inquiry is. Here then is another kind of dialectical shift where the expert consultation dialogue is embedded in the critical discussion in such a way that the one context of dialogue is functionally related to the other. The expert consultation dialogue actually improves the level of the critical discussion of the issue by bringing in all sorts of relevant "facts" or scientific findings. Here there is a shift or mixture of the two dialogues, but it is a positive or *licit shift*, as opposed to an illicit shift. Fallacies tend to be associated with illicit shifts.

Emotional Commitment and Bias

The methods of collecting data in an inquiry generally have to be *reproducible*, meaning that another investigator should, ideally, be able to go through the same procedure and get the same result. But arguments in a critical discussion depend on a participant's point of view. It is therefore harder to eliminate bias in a critical discussion than in an inquiry. And by the same token, it is more important for participants in a critical discussion to pay attention to the likely possibility that their opponents may be biased, one way or the other.

But bias is not inherently bad in a critical discussion. Ideally, critical doubt is the absence of commitment for or against a thesis at issue. But commitment, either for or against a point of view, is not, by itself, an indication that a fallacy has been committed. Bias becomes critically suspect when it is inappropriate, or hardens into dogmatism, as, for example, when there has been an illicit shift from a critical discussion to a quarrel.

One of the principal criticisms of the critical discussion as a method of rational argumentation that can lead toward knowledge or truth of a matter being discussed is that it is emotional. Any good critical discussion of an issue depends on strong partisan argumentation on both sides, based on commitment to defend one's side of a controversy. This kind of argumentation, however, seems to work best when it is based on the passionate conviction of the participants. But then inevitably they will use arguments that depend on appeals to emotion. Does this not compromise the whole enterprise? How can there be "impartial reason" when appeals to emotion are not only allowed, but may even be the strongest arguments?

What should one say here? Can appeals to emotion be accommodated to the rationality of critical discussion as a type of dialogue that leads towards truth? Or must all appeals to emotion, of any sort, be rigidly excluded from rational argumentation?

The approach taken by this monograph is that appeals to emotion can be admitted in a critical discussion and are legitimate in other contexts of dialogue as well, but they need to be evaluated by boundary conditions that rule them as fallacious in some cases. The thesis advocated here is not only that appeals to emotion can be allowed, but that certain types of arguments based on appeals to emotions can be very positive and contribute fundamentally to the goals of a critical discussion. At the same time, the negative thesis is advocated that these emotional arguments can be used fallaciously in particular uses so that they go contrary to the proper and legitimate goals of a critical discussion or other type of dialogue that participants are supposed to be engaged in. By learning to distinguish between the fallacious and nonfallacious uses of these arguments based on emotions, it is possible to use them constructively without falling into fallacies and errors.

The goal of a critical discussion is the resolution of a conflict of opinions. And the procedural rules of a critical discussion oblige the participants to cooperate in making moves that are appropriate for the given stage of the discussion. But what guarantee is there that a critical discussion will uncover knowledge or lead toward the truth of the matter being discussed? And failing any guarantee of this sort, what use is a critical discussion as a kind of rational argument?

The answer proposed here is that a good critical discussion can prepare the way for the reception of truth or knowledge through the discovery of bias and fallacies. In a successful critical discussion an arguer's position can be articulated, clarified, and deepened. The basis of an arguer's commitments can be brought to light when he or she replaces unquestioned presumptions with reasoned arguments. Through the maieutic function of dialogue, the arguer can come to a better understanding of where he or she stands on an issue of discussion. And this deepening of understanding both requires and is accompanied by insight into the opposed point of view. This understanding is not, in itself, a kind of truth or hard knowledge. It would be dogmatic, or perhaps Platonic, to portray it as such. It results from argument based on presumptions that are not known for certain to be true or false. But this is a kind of reasoning that is useful in the practical affairs of life and that can also be extremely useful in expediting the discovery of knowledge, a kind of reasoning with a critical perspective that takes the hard edges off a dogmatic attitude. A dogmatic arguer tends to go to excesses; by responding uncritically to emotional appeals and by committing fallacies and faults in argumentation, this arguer is led down many blind alleys.

Because it is possible for a critical discussion to be embedded in an inquiry, the critical discussion can expedite the inquiry. In science, for example, the inquiry part is the official presentation of established results in a field. The discovery stage often has much more of the flavor of a critical discussion, where, for example, an investigator may have to use skills of critical judgment to differentiate between what can be proved and what must rest on presumptions that cannot be proved within the field of the inquiry. Hence the critical discussion can have a preparatory function in contributing to an inquiry.

Evaluating Appeals to Emotion

This book focuses on the four emotional arguments *ad*—the *ad baculum, ad misericordiam, ad hominem*, and *ad populum*—and tries to determine the conditions under which they are used correctly or incorrectly. By 'conditions' is meant normative requirements that would enable a critic to evaluate the argument as strong, weak, faulty, or fallacious and to have good grounds (evidence) for doing so.

Contrary to the common assumption that an argument based on emotion is not a rational (reasonable) argument, such an argument can be good and reasonable insofar as 'good' or 'reasonable' argument is that which contributes to the proper goals of the dialogue. The critical discussion is one of the most common and important types of dialogue in which arguments based on emotional appeals are used in everyday conversations.

Arguments based on emotional appeal can be good in a critical discussion because they can link an argument to an arguer's so-called dark-side commitments on an issue. These are commitments that reflect the arguer's deeper, underlying convictions or position. Typically, they are not known to the arguer in any explicit form; they could be described as "veiled" or "deeper" commitments that the arguer is only aware of as "gut feelings." A critical discussion reveals these implicit commitments to an arguer.

Another contention brought out by many of the case studies in this book is that most argumentation in everyday conversation is based on presumptive reasoning. The issue to be argued out is a conflict of opinions, one not determined by appeal to existing or available knowledge. This feature is both good and bad, as far as appeals to emotion are concerned.

Appeals to emotion can be useful, and correct, in argumentation because they can steer an arguer toward a resolution of the conflict of opinion by burden of proof and weight of presumption. Gut feelings may reflect an arguer's experience, wisdom, or better instincts. As such, the arguments based on them are not decisive or absolute, but they can steer the line of argument in a favorable direction.

On the other hand, appeals to emotion can be a hindrance, blocking the proper goals of a critical discussion. Every participant in a critical discussion has a bias toward his or her own point of view. Bias is not, in itself, bad or obstructive in argumentative dialogue, but it might be so called when one's bias is so heavily favored that critical perspective in an argument is lost. Appeals to emotion, as appeals to an existing bias, can become a hindrance when they upset the proper balance steering an argument, and the argument goes too far to one side.

Bias is also associated, in severe cases, with fallacies, especially the four emotional fallacies studied in this book. Bias is not itself a fallacy, it is contended, but these four fallacies are associated with a hardening of bias. This hardened bias is in turn associated with the suppression of legitimate critical doubts concerning one's own point of view, and with illicit dialectical shifts from one context of dialogue to another.

To sum up, arguments that are based on appeals to emotion, of the types analyzed in this book, have a good side and a bad side. These arguments are good when they open up new and valuable lines of argumentation, prompting critical questioning that steers the argument in a constructive direction. From a pragmatic point of view, when used correctly such arguments in a dialogue open up deep feelings about what is important and right and steer the argument along by according important presumptions the weight they deserve. On the bad side, accompanying any line of presumptive reasoning is a natural tendency to jump to conclusions that favor one's own interests or bias too quickly. This lack of critical balance needs to be offset by an awareness of fallacies and the tendency to bias.

Two Reservations

In setting out the scope of this project, two careful reservations must be made. One is that this book is not meant to be an investigation of the psychology of the emotions but a normative analysis of the conditions under which appeals to emotion are used correctly or incorrectly in argumentation. The other reservation is that this book is not meant to provide an analysis of the concept of relevance, even though it must rest on some assumptions about relevance as a category of argumentation.

The four fallacies studied are all appeals to emotion or, in the case of the *ad hominem*, strongly associated with heightened personal emotion in argumentation. Defining them as types of arguments involves essential reference to emotions like fear or pity. Even so, the project here is a logical, normative, critical, and linguistic one, and not a treatise in empirical psychology, even though it is a practical investigation that is meant to have implications for empirical studies of argumentation by social scientists. Hence no definitions for emotions like fear will be ventured; nor will the text attempt to define in behavioral terms exactly what constitutes a threat to a person. These questions remain open to investigation by social scientists. Few uncritical assumptions will be made about them, and when assumptions are made, they will be defined clearly, so that they are left open to possible revisions by those qualified to do so. The aim will be to assist, not hinder, investigations of the emotions by the empirical methods of the social sciences.

The four *ad* fallacies of appeal to emotions have most often been classified by writers of logic textbooks as failures of relevance in argument. For example, Whately (see pages 6–7) categorized all six of the *ad* fallacies as species of the Aristotelian *ignoratio elenchi*, literally 'ignorance of refutation,' but usually translated as 'irrelevance,' 'red herring,' 'missing the point,' or 'ignoring the issue.' However, the concept of relevance has never been defined in a clear or useful way in the field of argumentation. It has become a kind of "wastebasket" category: if you can't figure out why an argument or fallacy is wrong, call it a "fallacy of relevance." Hamblin (1970, 31) notes the practice of treating *ignoratio elenchi* as a "rag-bag" category.

Currently a major research project is underway, undertaken by Professors Frans van Eemeren and Rob Grootendorst, in collaboration with the author, to provide an analysis of the concept of dialectical relevance that will be useful in the normative evaluation of argumentative discourse. In this project, relevance is characterized as a dialectical concept. A speech act is said to be relevant in the context of a critical discussion if it "fits into" that context; its relevance is determined by the initial conflict of opinions in the critical discussion, by the stage of argumentation the discussion is in, and by the localized context of speech acts and argumentation schemes.

The present project on emotional fallacies cannot supplant this project on relevance, but the intent of the author is for the two projects to complement and contribute to each other. Hence, in the present book, when relevance is referred to, it is the place and function of a speech act in a context of argumentative dialogue that will be meant.

2

PRESUMPTIVE REASONING

According to Plato, genuine knowledge is of the true and unchanging, whereas opinion is constantly shifting, saying one thing and then the opposite. Opinion-based argument is the domain of the sophists, arguers who seek fees instead of the truth.[1] The Platonic derogation of all opinion-based reasoning continues to exert a firm hold, reinforced by Descartes's insistence that genuine knowledge must be based on indubitable premises.

In fairness to Descartes, it should be acknowledged that he did not derogate all opinion-based or probabilistic reasoning. But such reasoning, for Descartes, did not qualify as genuine knowledge. Moreover, both Plato

---

1. Socrates was particularly hard on popular opinion as a source of premises for argumentation, putting down the opinions of "the many" as constantly changing due to the emotions that shift this way and that.

and Descartes were right to take the position that in any conflict between knowledge and opinion, knowledge should get priority.

Unfortunately, however, when Plato relegated opinion-based reasoning to a lower rank than knowledge-based reasoning, this cast a pall over all opinion-based reasoning for subsequent generations of logicians, who presumed that it was not worth systematically analyzing as part of the logic curriculum.

This chapter will break this Platonic hold by examining some traditional informal fallacies. Through the analysis of the important kinds of argumentation shown to be involved in these so-called fallacies, it will be revealed how presumptive opinion-based reasoning can be a legitimate framework for studying fallacies and other logical errors. The traditional criticism that opinion-based reasoning is subjective argumentation that continues indefinitely with no resolution will be offset by the twin concepts of presumption and burden of proof. The function of burden of proof is to support a presumption well enough to allow an argument to arrive at a provisional, but well-reasoned, conclusion. This concept of presumptive reasoning presupposes a goal-directed probative dialogue framework in which two participants interact by asking each other questions, and by drawing conclusions from the other party's premises. The goal is to prove something according to a given standard of proof.

Interactive reasoning and its arguments in the context of dialogue can be evaluated as good or bad on the basis of a critical understanding of the traditional informal fallacies. The fallacies represent not only the bad kinds of arguments, or faults of dialogue argumentation; they also, curiously, underlie structures of interactive reasoning (argumentation schemes) that in many cases can be evaluated as correct uses of argument.

Presumptive Reasoning and Knowledge

Logical inference has too often been taken in the past to be exclusively about proof, in the sense of "knockdown" irrefutable argument. Since Plato's all too successful attack on the sophists for dealing with opinions that cannot be proved as "true and certain knowledge," opinion has been regarded with deep suspicion as unfit territory for philosophers. In many ways, however, this old view is strongly at odds with the dialectical concept of philosophy and the value of humanities generally as fields of study based on interactive argumentation that reasonably examines both sides on controversial issues where conflicts of opinion are perfectly natural and where the "factual" basis for scientific inquiry is inadequate.

The Platonic view is that reasoned dialogue can never be a method that works toward or results in genuine knowledge as long as it is opinion-based. The popular presumption of the times is that only science is built on a solid foundation and is therefore an objective contribution to the advancement of knowledge. The humanities, based as they are on methods of reasoning in dialogue, are taken to be inherently subjective, and therefore second-rate as serious academic pursuits for the best minds. Can reasoning in dialogue truly be a logical way of arriving at conclusions that contribute to knowledge? If so, it is unclear by what exact mechanism it works. And when it does appear to work, it is unclear why or how its conclusions make a real contribution to knowledge. Nevertheless, in this century a heavy burden of proof has grown against the conclusion that it does.

One danger of dialogue, admittedly, is that it can go on and on with no conclusion. This problem can be solved by introducing the device of burden of proof. But burden of proof is based on presumption, and presumption is not knowledge. The danger of making interactive reasoning practical is that it appears to become subjective. Is there any escape from this predicament? This is a hard question—not one that any single book or article could hope to resolve completely. But in this chapter, a way out of the predicament will be proposed.

The danger of interactive reasoning as a method of contributing to knowledge is that the two participants in dialogue each have separate knowledge-bases or positions. Successful refutation by one party, therefore, yields disproof relative only to the other's knowledge base, position, or stock of commitments. If the two knowledge bases are different, what guarantee is there that the prevailing party's knowledge base is "genuine" or "real" and that the conclusion "proved" is reliable as an item of knowledge? Perhaps not much, for as examples of the *argumentum ad verecundiam* show, even an expert can be wrong!

So it seems that presumption is one thing, but proof is another. Even "reasoned" presumption by question-reply dialogue that adheres to a set of rules of procedure could typically fail to yield reliable enough conclusions to be properly called "knowledge." The pitfalls of parliamentary debate and other institutional forms of dialogue, like legal argumentation, are well known. These dialogues can become terribly confrontational and emotional and can encourage fallacies and blunders of the worst sort.

Abundant evidence of these pitfalls has accrued from the traditional study of fallacies.

The rational conviction of a respondent in dialogue is said to be secured if the proponent's argument has met the burden of proof appropriate for the context of dialogue. But it has often rightly been doubted by some philosophers whether a burden of proof of "absolute certainty" is ever appropriate for settling discussion of a subject (see Rescher 1977, 1988). It can be argued that the very best one can hope for in discussing any subject where genuine doubt can arise is to set the burden of proof "beyond reasonable doubt." According to this line of argument, a conclusion of a critical discussion can be relatively rational, to a greater or lesser degree, but it will always fall short of perfect conviction, free from all reasonable doubt. This line of argument could be called the *fallibilist interpretation* of critical discussion as a way of arriving at conclusions.

The concepts of burden of proof and presumptive reasoning only make sense in relation to the concept of an argument as a balance with weights on each side. A shift in the burden of proof means that as more weight is placed on one side, the other side becomes lighter (Walton 1988). Therefore, what is proved in argument is always relative to what the other side has proved. Criticisms have the purpose of shifting this burden, not of fixing it for all time.

Burden of proof is often practically linked to the problem of reasonably deciding on a course of action in an uncertain situation. In a rapidly changing or complex situation, certain knowledge or even probable knowledge may not be obtainable as rapidly as decisions are required. Often, instead, an agent must act on uncertain presumptions about what can reasonably or normally be expected to happen in a given situation, based on usual expectations, customary routines, and commonsense understanding of institutions, functions, and familiar sequences of actions.

Presumptive reasoning is appropriate where reasoned conviction is the best one can aspire to. This reasoning is based on shifts in a reasoner's commitments in reasoned dialogue on issues that are controversial or practical; the process yields a conviction despite lack of access to deductive certainty or even probability. Such reasoning is inherently fragile and therefore follows the conservative principle of fallible reasoning: a chain of argumentation is only as strong as its weakest link (Rescher 1976). To wit, a proof in interactive reasoning is only as reliable as the knowledge base or position of the respondent from which it was inferred. This makes such a conclusion fairly reliable (in many cases) if the respondent is an expert in the domain of the argument. But in other cases it leaves open the legitimate critical question whether the conclusion can rightly be said to be reliable knowledge.

However, rules of presumptive reasoning are not only applicable to expert systems reasoning and to reasoning about expert advice generally. They can also be applied to dialogue reasoning where two participants are involved in question-reply interaction governed by a set of rules of critical discussion for deriving conclusions in argumentation. In this case the set of premises involved in reasoning does not need to be a knowledge base. It can be a set of propositions or position base that represents the concessions or accepted commitments of each separate participant in the argument. In the basic case of a conflict of opinions, there are two sides, the *pro* and the *con*.

An arguer's position, or commitment set, represents the total of his or her set of commitments at any particular point in a sequence of reasoned question-answer dialogue. Burden of proof should be set before the sequence is initiated, at the opening stage ideally, but local weight of presumption can be shifted during the course of a dialogue. Indeed, reasoned dialogue is essentially a sequence of shifting presumptions. Therefore, whether argument in interactive reasoning can produce useful and insightful conclusions depends on the use of the rules that determine correct and incorrect shifts of presumption in relation to a burden of proof in a dialogue. However, following these rules doesn't guarantee that the outcome of a dialogue will establish the absolute truth, or even that the side that makes the strongest case for its contention will have the last word on the subject.

Presumption Defaults and Fallacies

It is often presumed in logic textbook treatments of fallacies that presumptive reasoning that arrives at a conclusion by intelligent guesswork rather than by scientific knowledge or thorough inquiry can be automatically classified as fallacious. Engel (1976, 70) offers as an example of "the fallacy of hasty generalization" the conjecture made by Sherlock Holmes upon first meeting Dr. Watson. Seeing Watson's deep tan, military bearing, injured arm, and haggard face, Holmes concludes that Watson must have returned recently from military service in Afghanistan. Holmes is right, but Engel claims that he has committed a fallacy, because his conjecture is "founded on insufficient evidence." But insufficient for what? Because Holmes puts forward a defeasible presumption based on signs, a conjecture that can then be verified or refuted by Watson, has he committed a fallacy? Only if Holmes's conclusion is meant to be taken in a rigid and unqualifiable, nondefeasible way.

Many rules and inferences used in everyday argumentation are of a qualified or defeasible type that cannot be literally or automatically applied to all cases. A proposition can be put forward in argumentation as a *generic statement*; it is meant to be applied in particular cases in a way that is inherently open to exceptions. For example, "Birds fly" may be put forward not as an exceptionless qualification, but as a generic statement that is meant to be compatible with the existence of dodos and kiwis (see Reiter 1987). It need not be understood to imply that anything, if it is a bird, must fly. As a generic statement it can accommodate exceptions in some cases—for example, in the case of a bird with an injured wing.

It would seem quite plausible as well that overlooking the generic nature of some propositions put forward in argumentation could lead to errors and confusions, even to the occurrence of fallacies. These problems could easily arise from shifts of context where differing standards of burden of proof are appropriate. To require that an opponent's thesis be proved *without exception* is a high standard of proof that may be inappropriate and unfair in many contexts of dialogue. The insinuation of such a requirement into the dialogue could be the basis of clever sophistical tactics in argumentation. According to the traditions of logic, there are indeed fallacies associated with just such tactics. But unfortunately, at least for the time being, the standard treatment of this type of fallacy seems to be confusing and contradictory.

According to Hamblin (1970, 28) the fallacy of *secundum quid*, meaning "in a certain respect"—in Greek, *para to pē*—refers to failures of correct argumentation following from neglect of qualifications that should be attached to a term or generalization. This seems clear enough, but Hamblin went on to criticize the standard treatment of *secundum quid* that has covered similar, or perhaps even the same errors, under the headings "hasty generalization" and "fallacy of accident." The problem is that neither of these other two purported fallacies is treated in a clear or coherent way by the logic textbooks. The result is a variable and confusing treatment of this whole area.

According to Copi (1986, 99), the fallacy of accident "consists in applying a general rule to a particular case whose 'accidental' circumstances render the rule inapplicable." Unfortunately, this description is virtually identical to that traditionally given by most textbooks to the fallacy of *secundum quid*. Hamblin (1970, 30) conjectured that this unfortunate reclassification was due to a careless reading of DeMorgan by writers of textbooks.[2]

Even more confusing and unfortunate is the intrusion of the term "fallacy of accident" into this territory. Originally, this so-called fallacy came from Aristotle's theory of essential properties and was supposed to be the fallacy of "taking an accidental property to be an essential one." (Hamblin, 1970, 27). But nowadays nobody takes Aristotle's theory of essential properties seriously, either as suitable subject matter for logic textbooks or as an established branch of logic.

The term "hasty generalization" (overgeneralization, sweeping generalization, etc.) is confusing also because it is often applied to errors of insufficient statistics (too small samples) by the textbooks, as well as to defaults of presumptive inferences by virtue of application to exceptional cases. But these failures of inductive and presumptive reasoning are really two separate types of errors that become confusing when uncritically lumped together.

Fearnside and Holther (1959, 13) describe the fallacy of hasty generalization as the fallacy of "generalization from too few cases . . . , which statistical science shows to be insufficient." Engel (1976, 66) describes the fallacy of sweeping generalization as "committed when a general rule is applied to a specific case to which the rule is not applicable because of the special features of that case." Whereas Fearnside and Holther's fallacy is clearly meant to be an inductive error of too small sampling, Engel's fallacy is the same one that was traditionally, and less confusingly called *secundum quid*.

For the best classification (1) keep the fallacy of hasty (sweeping, etc.) generalization as another name for "insufficient statistics," an inductive fallacy (see Walton 1989a, 206), (2) get rid of "the fallacy of accident" altogether, and (3) keep the name *secundum quid*, which could be called in English the fallacy of insensitivity to special cases.

Failure to be aware of shifts of presumption in exchanges in dialogue lies at the basis of many fallacies and undetected errors and lapses of effective argumentation. For example, failure to react aggressively in countering an

2. Hamblin refers to the section on fallacies in the textbook of DeMorgan (1847).

ad hominem attack or in replying to a loaded or complex question can be a serious failure in argumentation. By failing to insist that the burden of presumption should be on the attacker to support allegations, a respondent may appear to have gone onto the defensive too easily and may therefore be open to suspicion. Such a respondent overlooks the best defense, that of shifting the burden of proof back where it properly belongs: on his opponent's side of the dialogue.

Whately (1846, 114) pointed out this error: "Let any one imagine a perfectly unsupported accusation of some offense to be brought against himself; and . . . instead of replying . . . by a simple denial, and a defiance of his accuser to prove the charge . . . , [he] takes on himself the burden of proving his own innocence." Whately (113) compared this tactical failure to a case where troops in a fort strong enough to repel any attack, ignorant of their advantage, sally forth into an open field to meet the enemy and are decisively defeated. According to Whately's analysis (discussed in "Historical Background," Chapter 1), the root of the problem is a failure to be aware of the shifting of a weight of presumption in a dialogue.

Thus, failures of presumptive reasoning of various kinds are associated with important kinds of fallacies. But it does not follow that presumptive reasoning is always fallacious, because such reasoning is not based on firm evidence that proceeds from extensive or definitive inquiry.

Another traditional fallacy requiring radical reappraisal once the notion of presumption is taken seriously in argumentation is the *ad populum*. Copi (1986, 96) states that emotional appeals "to the gallery" are often used as sophistical tactics to cover up a lack of evidence in advertising and propaganda. A belief is not necessarily true just because "everyone knows it" (105). Appeals to popular opinion may be erroneous, or they may merely be deceptive, used to silence opposition in ways that are at odds with critical discussion of an issue.

But surely it is reasonable to take the weight of public opinion into account in formulating an initial weight of presumption, in favor of or against a particular proposition, when, for example, setting reasonable burden of proof on a controversial issue of public affairs. At least, it can be conceived as a reasonable way of proceeding in some contexts of dialogue where the argumentation should be based on presumptive reasoning.

Take the case of an epideictic speech as a context of dialogue, for example, where the goal of the speaker is to emotionally instill solidarity and patriotic feeling in his compatriots at a special, public ceremonial occasion.[3] Is the appeal to popular sentiment out of place here? Is it a logical fallacy wherever it occurs? Clearly the answer is "No." Popular emotional appeal should only be evaluated as fallacious in specific situations where it has been used erroneously or sophistically to go against the proper goals of the dialogue, for example, by covering up a lack of appropriate proof or evidence in a critical discussion.

Argumentum Ad Ignorantiam

Because presumptive inference is most useful where there is lack of conclusive evidence to decide an issue, it is a kind of reasoning closely associated with lack-of-knowledge argumentation, that is, with the *argumentum ad ignorantiam*, a kind of argument traditionally classified as a fallacy. In many cases, however, the argument to ignorance is not a fallacy at all. It is a reasonable use of presumptive inference in argumentation.

Legal argumentation is full of reasonable uses of *ad ignorantiam* argumentation. It is a legal principle concerning wills and inheritances, for example, that if a person has not been heard from for seven years, there is a presumption of death (see Ilbert 1960; Degnan 1973). This is a kind of *ad ignorantiam* reasoning—if Jones has not been heard from for over seven years and there is no evidence that he is still alive, then a legal inference is put into place with the conclusion that Jones is presumed (for purposes of his estate) to be dead. Of course, such a presumption is not meant to exclude all possibilities that Jones is alive. If, in fact, Jones turns up, the presumption will be cancelled. The function of the inference is to shift the burden of proof onto any claimant that Jones is alive to produce sufficient evidence to rebut the contrary presumption.

Whether an *ad ignorantiam* argument is reasonable or fallacious in a particular case depends on the standard of burden of proof for the dialogue context of the argument. Normally, *ad ignorantiam* arguments can function quite successfully as shifters of a weight of presumption in a dialogue where there is a relatively even balance of presumption between the two contending sides and the preponderance of credibility can easily be shifted from one side to the other. But if the weight or force of the *ad ignorantiam*

3. See Perelman and Olbrechts-Tyteca 1969 (46–51). See also the "Presumptive Inferences" section of Chapter 3.

is radically overestimated—for example, if it is taken as a conclusive demonstration of the proposition in question—it can easily become a fallacy (Robinson 1971).

A case illustrating both the use of the *ad ignorantiam* in legal argumentation and this shift to a standard of proof appropriate for a different type of dialogue is given by Copi (1982, 102).

Case 2.1

> It is sometimes maintained that the *argumentum ad hominem* (abusive) is not fallacious when used in a court of law in an attempt to impeach the testimony of a witness. True enough, doubt can be cast upon a witness's testimony if it can be shown that that witness is a chronic liar and perjurer. Where that can be shown, it certainly reduces the credibility of the testimony offered. But if one goes on to infer that the witness's testimony establishes the falsehood of that to which the witness testifies, instead of concluding merely that the testimony does not establish its truth, then the reasoning is fallacious, being an *argumentum ad ignorantiam*. Such errors are more common than one might think.

This case also shows, interestingly, how the *ad hominem* argument is often used as a species of *ad ignorantiam* argumentation to cast doubt upon the credibility of a witness or other source of presumptively based testimony or opinion.

As long as it is clear that the *ad hominem* questioning of the veracity of the witness is only being used to cast doubt upon the presumption that the witness is giving honest testimony (see Graham 1977)—and assuming, as Copi puts it, that "the witness is a chronic liar and perjurer"—then the use of this *ad ignorantiam* line of argumentation to shift a burden of presumption against the credibility of the witness's testimony could be quite reasonable. Where it goes wrong, according to Copi's account, or becomes a fallacious argument, is in "establishing the falsehood" of the witness's testimony. For establishing falsehood is a high standard of burden of proof, appropriate perhaps for a scientific inquiry or demonstration where the facts have been carefully established, but highly inappropriate when dealing with the cross-examination of a witness whose testimony is teetering on the borderline of credibility or noncredibility. The fallacy then is to take the *ad ignorantiam* argument for something it is not properly meant to be in this context—a knockdown, conclusive, demonstrative establishment of truth or falsehood, instead of a weak presumption-shifter.

The importance of this shift in the context of dialogue for evaluating the *ad ignorantiam* is well brought out by Copi's next example, designed to show that this type of argument can be reasonable in a context of dialogue where it is not based totally on ignorance but partly on the findings of an investigation. Copi (1982, 102) asks us to consider the following case.

Case 2.2

> A serious security investigation fails to establish that Mr. X is a foreign agent. It would be wrong to conclude that their research has left us ignorant. It has rather established that Mr. X is not a foreign agent.

In this case, absence of a finding can function as positive evidence or argumentation that something does not exist or is not there to be found. Copi writes that "the proof here is not based on ignorance but on our knowledge that if it had occurred it would be known" (102). To the extent that this is true, however, the argument in case 2.2 is not an argument from ignorance at all but a positive argument based on knowledge or evidence that has accrued as the result of an inquiry.

Whether an *argumentum ad ignorantiam* of the type illustrated in case 2.2 is reasonable or fallacious then depends on the context of dialogue. Once some standard of burden of proof is set as reasonable for the context of dialogue, then the progress of the dialogue must be examined. In the case of an inquiry, one needs to evaluate how far along the investigation has gone and how thoroughly it has established argumentation to meet that burden of proof (see Glucksberg and McCloskey 1981).

The *argumentum ad ignorantiam* is not, generally speaking, a fallacy, although it is a kind of argument that can be fallaciously used in some cases. Rather, the *ad ignorantiam* is a principle or type of argumentation that reflects the shifting of presumption from one side of a dialogue to the other according to the requirements of burden of proof for that type of dialogue.

The intimate conceptual connection between the *argumentum ad ignorantiam* and the use of presumption as a device to facilitate action or advance a dialogue by shifting a burden of response in argumentation is graphically brought out in the following case from Walton (1989a, 272) of a memo sent to a university faculty from the library.

Case 2.3

> The Library Staff is reviewing the policy of keeping old university exams on file for student use. It has been found that the majority of exams are more than 10 years old. For some departments, we have only 2–3 exams. Please discuss this with other members of your department. Please report back to me by September 22, 1986. *If no response has been received by this date, it will be assumed that you are in favor of disposing of the practice of keeping university exams.*

Using the negative conditional *ad ignorantiam* device, presuming in case 2.3 that no response indicates acceptance of the proposal, could be quite legitimate (depending on the particulars of the case, whether the deadline is reasonable, and so forth). It is a way of soliciting possible objections without having to call a meeting or soliciting responses from individual faculty. If numerous or serious objections came in as responses to the memo, the library could put the item on the agenda of an upcoming meeting of the faculty library committee. But the use of presumption in the memo could possibly function as a facilitating device for saving a meeting and allowing prompt action.

Here the context of dialogue is not that of an inquiry or investigation into a subject but rather that of a solicitation of opinions in a policy discussion prompted by a need to consider action on a particular question. The effect is to shift the weight of presumption onto anyone who has serious objections to the proposed action.

Hard and Soft Evidence

In many cases, a conclusion is accepted on the basis of presumptions in the absence of objective knowledge. Even though this reasoning is based on presumption, it can rightly be called "evidence" in some cases. It is appropriate to call this presumption-based reason for accepting a conclusion *soft evidence*, as opposed to knowledge-based *hard evidence*, which is established through inquiry based on objective facts or hypotheses that can be verified by testing in a reproducible way. Soft evidence is defeasible, and based on a person's say-so. It therefore depends on the accuracy and honesty of the person who reported it and who will stand behind it, cooperatively submitting to critical questioning concerning it.

Eyewitness testimony is one form of soft evidence. Appealing to the opinion of an expert is another. Both of these can easily go wrong. But Loftus (1979) has shown that in the case of eyewitness testimony there are techniques for confirming or rebutting conclusions based on soft evidence. One, of course, is careful cross-examination, testing the respondent's presumptions and assertions for consistency and plausibility.[4]

Sometimes soft evidence is produced by intelligent but tentative guesswork—a defeasible reasoning that can nevertheless be very useful. Presumptions may be based on indications or signs that are not firm evidence but give a basis for conjecture. A presumptive inference of this type can be strengthened by the finding of additional, positive signs.

Case 2.4

> During a visit to the Netherlands, Karen and Doug were cycling past a large complex of buildings they could see across the street from the bicycle path near the Central Station in Leiden. Doug saw a sign on one of the gates to a smaller adjacent building; it read "Ambulancieren." Doug said to Karen: "I think it must be a hospital. The sign over there looks like it refers to ambulances."

In this case, the building carried no mark that would reveal its function to Karen and Doug other than the "Ambulancieren" sign. Doug did not know this word in Dutch, but it seemed likely that it meant what the similar English word means. Thus Doug's conjecture was really a guess, but it seemed, in the situation, like quite a reasonable assumption.

Case 2.5

> As they cycled further past the large complex of buildings in Leiden, proceeding towards the Central Station, Karen replied: "I think you are right, because yesterday in Leidschendam, you remem-

---

4. See Ilbert 1960 and Degnan 1973 on the use of eyewitness testimony as legal evidence.

ber, we went past a similar building complex, and it had a sign on it saying that it was a hospital."

In the second case above, some additional evidence to back up the conjecture of the first case is brought forward by the other participant in the discussion.

Doug's initial conjecture was based on some "evidence," or at least a sign that appeared to be evidence (perhaps soft evidence), even though his interpretation of the sign could easily have been mistaken. But just for the amusement of speculation, he made a guess about the meaning of the sign. Karen could have rejected this guess as dubious, but instead she provided a little more soft evidence (that could easily be mistaken as well) to back up Doug's conjecture. As this new indication is brought in by Karen, it tends to bolster Doug's conjecture, making it a little more plausible as a presumption.

In these cases, Doug's opinion could be called more a conjecture or a guess than a presumption, although it is a kind of presumption to the extent that Doug has advanced it for Karen's approval or acceptance, inviting her, as it were, to provide counterindications if she can. But the opinion would function more overtly as a working presumption if there arose a need to act on it.

Suppose that Doug became violently ill and clearly needed immediate medical attention. Given the indications noted above, that the complex of buildings across the street was a hospital, and the absence of evidence of any other medical facilities in the area or sources of medical assistance quickly available, it could be quite reasonable for Karen to act on the presumption that this complex of buildings was a hospital.

Like the evidence cited in these cases, eyewitness testimony is also a common basis for argumentation. But is it a kind of empirical evidence, or is it merely subjective say-so, that is, a form of plausible or implausible presumption? If a witness, John Smith testifies that he saw Rita Jones at the Central Station in Leiden at 7:30 p.m. on 15 September 1989, then John Smith has made an empirical claim or attestation. Whether Rita was there or not at the time is an empirical question. But in relation to some disputed question about controversial events that took place near Leiden on that night, the parties discussing that question are drawing out as "evidence" a particular proposition as the conclusion of an inference. For example, such an inference might take the following form:

P1 John Smith says he saw Rita Jones in Leiden at the time specified.

P2 John Smith is a reliable witness.

P3 There is no evidence that Rita Jones was not in Leiden at the time specified.

C Rita Jones was in Leiden at the time specified.

Premise P1 is empirical in nature. However, premise P2 depends on whether or not John Smith may be judged to be a reliable source of information on the question being discussed. If it can be shown that Smith lied on previous occasions, that would be relevant evidence to rebut P2 as a reasonable presumption. P2 generally depends on Smith's character and in particular on Smith's reputation for veracity. It might also be relevant to an intelligent assessment of P2 to know if Smith has a lot to gain by claiming he saw Rita in Leiden at that time. In short, P2 is not (straightforwardly, at any rate) an empirical premise that should be supported or rebutted in the same way that P1 should be.

Moreover, the explicit insertion of P3 as a premise highlights the presumptive nature of the conclusion C. If evidence does come forward that rebuts P3, it will have a negative impact on the acceptance of P2, and this effect will, in turn, reasonably undermine C as a correct conclusion to be inferred.

It will help clarify the reasoning in this evidential situation to note that C is drawn by inference from premises that depend partly on empirical evidence and partly on the reliability of a source of testimony. Relying on a subjective source is different from using empirical evidence.

A useful distinction might be made between "soft" and "hard" evidence as follows: Confirming or disconfirming a proposition by using empirical evidence (or mathematical calculations) can be characterized as the use of hard evidence in argumentation. In such a context, it is appropriate to speak of "knowledge" if the evidence is substantial enough. By contrast, to weigh a proposition as plausible or implausible on the basis of the testimony of an allegedly reliable source can be classified as soft evidence in argumentation.

This classification is controversial because some would say that what is called soft evidence above is not really "evidence" at all. One can see the point in this objection, for in a scientific inquiry, subjective eyewitness testimony might not be considered admissible evidence. But in other contexts of argumentation, for example in a legal trial, eyewitness testimony is classified as a very important kind of evidence.

In legal trials, witnesses are required to swear an oath before testifying, which exposes them to prosecution for perjury should they lie. The requirement that a witness should give firsthand knowledge of the facts instead of his opinions or conclusion is a restraint also intended ensure the veracity of testimony. This requirement is called the *opinion rule* (Degnan 1973, 908), according to which it is supposed to be the function of the jury to draw conclusions or form opinions. However, exceptions to the opinion rule are made in the case of one special class of witnesses—expert witnesses. An expert is allowed to venture a supposition, even where it is only a matter of opinion, in a case where the matter is so technical that the jury would have difficulty without the help of the expert (Degnan 1973, 908).

The different handling of expert and nonexpert testimony in legal argumentation reflects the underlying difference between these two kinds of testimonial evidence in general. Both kinds of argumentation assume that the source is competent and reliable. But in the case of the nonexpert source, "reliable" means "honest," of good moral character. In the case of the expert witness, however, reliability and competence include much more, such as the expert's level of skill, as attested to by his or her qualifications, attainments, and the evaluations of colleagues.

The argumentation scheme for nonexpert (eyewitness) testimony is the following, where *a* is a source of testimony and A is a proposition.

> *a* testifies to witnessing A as true.
>
> *a* is a reliable witness.
>
> There is no evidence that A is false.
>
> Therefore, A is true.

By contrast, the argumentation scheme for drawing a conclusion from expert testimony is more complex. What the third premise makes clear is that argumentation from testimony is a species of *argumentum ad ignorantiam* that works by shifting a weight of presumption.

Before presenting the argumentation scheme for expert testimony, it is useful to examine the traditional approach in logic to fallacies of appeal to authority in argumentation.

Argumentum Ad Verecundiam

According to the tradition of the logic textbooks, *argumentum ad verecundiam* is the fallacy of the erroneous use of expert opinion to persuade someone to accept a proposition in argumentation. Although appeal to the authority of expertise, as a reason for accepting a conclusion, has long had something of a bad reputation in logic and popular opinion, in recent times it has become more widely accepted as a pivotal influence on opinion in a problematic situation.

When objective knowledge is not available to a person facing a procedural decision, appeals to expert opinion can be a legitimate way of obtaining advice. On the other hand, such authority-based arguments can become questionable or fallacious when wrong inferences are made from the advice or when the respect such authority commands is misused to silence an adversary.

The phrase *argumentum ad verecundiam* in Latin means literally "the argument to modesty." John Locke, in his chapter "Of Wrong Assent, or Error," in his *Essay Concerning Human Understanding* ([1690], quoted in Chapter 1, pages 7–9), explains the meaning of this curious phrase very clearly. As Locke describes this tactic, the intent is for one person to "prevail on the assent" of another by citing the opinion of some third person who has "gained a name" so that it would seem a breach of modesty to doubt the opinion of this learned person. In other words, the tactic portrays the person who does not readily yield to the authority of this learned expert as an individual who is impudent, insolent, or immodest. The suggestion is that anyone who does not accept the authority of an expert and who refuses to yield to it in argumentation is being unreasonable and perhaps even irrational. Certainly such a person is not collaborating with the ideals of a reasonable discussion, because he is being an unfair arguer who is using the *argumentum ad verecundiam* to force the opposition into submission.

Locke's explanation of the *argumentum ad verecundiam* is particularly insightful in that it does not deny that appeal to the opinion of an expert can be a reasonable kind of argumentation, while at the same time it shows how such argumentation can be abused as a tactic to force someone to accept a proposition when the evidence actually given is not strong enough to merit such acceptance. So portrayed, the *argumentum ad verecundiam* is a misuse of the appeal to expert opinion in argumentation.

The textbooks generally characterize *argumentum ad verecundiam* as the fallacy of illicit appeal to authority in argumentation, but the term "authority" is ambiguous in this context and can have several distinct meanings. One meaning is that of 'administrative authority,' the right to exercise command over others or to make rulings binding on others through some sort of office or recognized position. Another meaning refers to expertise. Wilson's cognitive authority (his second meaning for authority) may be defined as a relationship between individuals where what one says carries weight or plausibility for the other within a given domain or field of expertise (1983, 13). DeGeorge (1985, 13) distinguishes between an authority in a field of knowledge and an authority who occupies a position that carries rights or powers, not necessarily in virtue of expertise or knowledge.

Thus, broadly speaking, one can distinguish between two kinds of authority—administrative authority on the one hand and cognitive authority on the other—even though these two kinds of authority may be combined in the same individual in some cases. For example, a physician may be a cognitive authority, that is, an expert in the field of medicine, and an administrative authority whose standing as a licensed physician makes his or her rulings authoritative and binding on some questions. Usually the textbooks in logic have cognitive authority in mind when they cite cases of the fallacious *argumentum ad verecundiam*, but it can be very easy to confuse cognitive authority with administrative authority, and one must be careful, then, to distinguish the two types of appeal to authority that can be involved. The confusion between these two types of authority has often resulted in the suggestion that any appeal to authority is inherently coercive and to be treated with suspicion. For example, the word "authoritarian" is often used in this sense to refer to some overbearing or overly forceful use of authority. It is easy to confuse this connotation of administrative authority with the idea of cognitive authority and conclude, therefore, that any kind of appeal to cognitive authority in argument should be greeted with suspicion or hostility. Because of this confusion, people have always suspected sources of authority of being inherently subjective, and appeals to authority of *not* being appeals to real knowledge, and therefore of no serious import in a scientific investigation. As Hamblin (1970, 43) pointed out, arguments from authority have been particularly disliked at some historical periods; indeed, it could be said that the rise of scientific method was to some extent, a reaction against the role that authority played as a form of argumentation in medieval philosophy and science.

What then, is the real place of authority and, in particular, the authority of expertise or cognitive authority in recent argumentation? In order to begin to analyze this question, it is necessary to start with the presumption that a clear distinction can be made between cognitive and administrative authority, even though these two kinds of authority may be expected to overlap significantly in many cases of the *argumentum ad verecundiam*.

There are three common errors that are found when expert opinions are cited in argumentation. The first concerns the relevance of the field of expertise. If an expert's opinion is cited on a problem or question in one field, but the expert's domain of knowledge is in another field, the argumentative value of the appeal to expert opinion may be dubious. For example, Smith may be a famous scientist with impeccable credentials in the field of biology, but if her opinion is being cited on a political matter, one should question whether any legitimate transfer can be made from the field of biology to the area of politics. It may be that Smith, like many academic specialists, has been kept very busy working at her area of expertise and has political views that are naive or perhaps not well informed. The problem here is to resist the 'halo effect' that experts have, for if someone is acknowledged as a legitimate expert in a field of knowledge, then that person often has a halo of authority that may carry over into anything he or she might say, even outside his or her area of expertise. Many fields of expert knowledge are narrow specialties, and attempting to transfer an expert's opinion outside this explicit domain can lead to fallacies and dubious claims (see Hoffman 1979).

The second problem with appeals to expertise in argumentation is vagueness. In some cases the name of the supposed expert who is consulted is not even mentioned, or perhaps the field of expertise in which this source is supposed to be cited is not mentioned either. For example, one often finds appeals to expertise prefaced only by the phrase "according to the experts. . . ." Any appeal to expert authority that is this vague may be so difficult to document that it would be an error to accord it any serious weight in argumentation. News media reports often use the phrase "according to the experts" without giving further documentation. These appeals are weak but could sometimes be backed up by citing legitimate expert sources that the news media used. In other cases, the use of this phrase may be an attempt to disguise the lack of any research, any effort to back up a claim by appeal to legitimate expert opinions (Shepherd and Goode 1977). In any event, unless the name or field of knowledge of the expert is given in some precise way that can be documented and followed up, it may be impossible to differentiate between the useless and the useful appeal to expert opinion in an argument. Unfortunately, these kinds of claims often go unchallenged, either because the audience is intimidated by the authority of an appeal to expertise or simply because nobody bothers to challenge the claim. Hence, weak appeals to expertise of this sort need to be challenged or further investigated if they are to stand up to critical scrutiny and carry any weight as reasonable arguments.

In the third kind of error the expert is identified but is simply not an authority in the relevant field of expertise. In cases like this, a source may have a certain credibility because he or she is famous or highly thought of generally, but if the claim is meant to be advanced on the basis of expertise, one should always question whether this person is a legitimate expert in the relevant field. The case where a show business celebrity recommends some commercial product should be familiar to all. In this case it is often not clear whether the argument is meant to be on the basis of expertise or not, but such arguments may often be taken as based on some kind of expertise, especially where claims are made about matters like nutrition, weight loss, and so forth, because the person's very appearance may be taken as testimony to an expertise or personal knowledge about the benefits of the product the celebrity is advocating. Failing proof of expertise, such claims should not be given credibility, and on reflection, it may be clear that there is no basis for such a presumption. In fact, the error here likely is that the person cited is no real expert at all (Weber 1981).

Argumentation Scheme for Appeal to Expert Opinion

The argumentation scheme for the appeal to expert opinion has three premises and a conclusion. The focal premise is that the individual cited is a genuine expert in a particular domain of knowledge. The second premise is that this individual asserts that a particular proposition is true. The third premise is that the proposition cited is within the specified domain of expertise. The conclusion is that the proposition in question may plausibly be taken to be true. The conclusion of this argumentation scheme is a presumption, meaning that it is something that can provisionally be accepted as plausibly true in the absence of direct evidence to the contrary. It also means that the appeal to expert opinion in argumentation, using the argumentation scheme above, is inherently subjective, in that it must give way before objective scientific evidence. It is not a substitute for an appeal to scientific or empirical evidence but only a useful kind of argumentation that comes into play when access to firm observational evidence is not available for reasons of cost, time, or convenience. Nevertheless, the appeal to expert opinion in argumentation can be evaluated as a correct or incorrect kind of argumentation in a particular case, using objective criteria. Any use of this argumentation scheme can be judged as weak, erroneous, insufficiently documented, or even fallacious, if one or more of the premises in the argumentation scheme is not backed up by good reasons when queried.

There are six critical questions that are appropriate when evaluating an appeal to expert opinion in argumentation. The first critical question is to ask whether the opinion advanced by the expert actually falls within the field of expertise of that expert.

The second critical question is whether the authority cited really is a legitimate expert in the field in question. This can be evaluated by looking at factors like degrees or professional qualifications this person may hold, by looking at the expert's track record, by asking whether the evaluation of colleagues supports this person's claim to expertise, and by looking at the evidence of this person's publications or other accomplishments in the field.

The third critical question concerns the degree of authoritativeness of the expert cited. If the question at issue is within a subfield of a field of expertise, then the individual who is an expert in this particular subfield will have greater authoritativeness than another person who is a legitimate expert in the general area but who is not an expert within the particular subfield. For example, supposing that Dr. Jones and Dr. Smith are both physicians, but Dr. Jones is a general practitioner and Dr. Smith is an expert in thyroid surgery. If I am deciding whether or not to have thyroid surgery, the opinions of Dr. Smith may be more authoritative than those of Dr. Jones (Weber 1981).

However, evaluating specific cases where relative authoritativeness of experts is at issue presents many problems. For example, often a senior scientist who is administrative head of an established institution may look to an outsider like an impressively credentialed expert, but to those "in the know" this individual may be more of an administrator than an authority on the latest developments in some field of research that is being investigated in the institution he represents.

Shepherd and Goode (1977) found that in reporting scientific developments, the news media tend to seek out the administrative head of an institute or faculty rather than a working researcher, who would often be a much more authoritative spokesperson on the latest developments. However, from a point of view of public credibility, it often seems more impressive to quote the head of an organization than to quote a mere working scientist in the lab. The problem here is that it is hard for a layperson who is not familiar with the field to judge who the real experts are.

The fourth critical question concerns disagreements among several qualified experts. While experts often agree about some things, they also often disagree about other things, and when this kind of disagreement occurs, it may be difficult for a layperson to arrive at a reasoned decision on whose word to accept. Usually the best way to deal with this kind of conflict is by a further questioning of the experts.

The classic case of this problem is the battle of the experts in the courtroom brought in by each side to make scientific pronouncements in support of that side's case (Hoffman 1979; Imwinkelried 1986). In this forum each attorney gets a chance to cross-examine the other side's experts and by a series of probing questions hopes to critically scrutinize the expert's opinion in search of weak points.

This cross-examination is worth emulating; it is often useful and even necessary for a layperson to ask questions of an expert, to ask for explanations, and to try to make sense of what the expert has said in relation to the layperson's problem. This process of dialogue involves practical reasoning and is often difficult because the layperson must grapple with technical information and terminology that is unfamiliar. Despite its difficulties, however, in many cases of ordinary decision making, it is necessary to tackle this problem if one is to make best use of the available information (Imwinkelried 1981).

The fifth critical question is whether there is objective, empirical, or scientific evidence available in relation to the problem at issue and, if so, whether the expert's opinion conforms to this available evidence. If scientific evidence is available, then this should generally be given preference to the say-so of an expert. However, it many cases, while some scientific evidence may be available in relation to a problem or question to be resolved, the evidence is not firm enough or clear enough to completely resolve the problem. In this case, then, it still may be useful to have the say-so of a qualified expert as partial guidance. Nevertheless, it should be required that the expert defend his position by explaining and dealing with the available objective evidence.

The sixth critical question concerns the interpretation of the expert's opinion by the layperson. To be useful, an expert opinion should be in a form that is clear and intelligible. Experts, however, often speak in technical jargon that resists translation into clear layman's term. And, being what they are, they may make all kinds of qualifications and exceptions for understanding the subtleties of a problem. Their advice itself may be qualified. If the expert's advice is to be genuinely useful, these qualifications should be carefully taken into account. Thus, in using an expert's advice, great care and attention should be given to the exact wording in which the advice is offered. With some appeals to expertise, however, meeting this requirement is a problem, particularly where the expert's opinion has not been quoted directly. In these cases it is necessary to question whether the expert's view has been presented accurately. Sometimes it is possible to check the original quotation and see whether the rendering of the expert's opinion is consistent with the words used originally to render his or her advice.

If an appeal to expertise takes the proper form of argumentation scheme as outlined above, then it shifts a burden of proof in favor of the proponent. However, asking any one of the six critical questions just listed can shift the burden of proof back onto that proponent to provide further evidence in support of his or her argument. But it is important to distinguish here between an appeal to expert opinion that is weak or lacks documentation and one that is fallacious.

Legal Uses of Expert Testimony

One institutionalized context where the appeal to expert opinion is accepted as a legitimate kind of argumentation is the use of testimony in the courts. Normally, when a witness gives testimony in court, that testimony is confined to firsthand knowledge of the facts, and a witness is discouraged from offering private opinions or conclusions. However, an expert witness is allowed to offer opinions or conclusions on a subject. This testimony is permissible in the courts because in some instances it is necessary not only to obtain evidence from specialists but also to have them interpret that evidence for a judge or jury who otherwise would have difficulty understanding it.

A good example is the expert judgment of physicians, which is often a necessary part of the evidence in criminal cases. Until recently, the standard for the use of expert evidence in the courts was based on the ruling of *Frye v. United States*, 1923. This ruling required that any technique or theory used in evidence must be "sufficiently established to have gained general acceptance in the particular field in which it belongs" (Imwinkelried 1981). But this ruling has often come under criticism for excluding newly developed scientific techniques. Such criticism has led to a recent liberalization of standards of expert testimony, which has given more power to experts in many cases.

An example cited by Imwinkelried (1986, 22) is that many courts now permit psychiatrists to testify that the psychological problems of an alleged child abuse victim constitute evidence that such abuse in fact occurred. This kind of testimony is based on the notion that children develop characteristic syndromes a psychiatrist or a psychologist can detect. Whether such syndromes really exist and what they are, however, is often a subject of controversy within the fields of psychiatry and psychology. Therefore, under the Frye ruling, this kind of evidence might not have been admissible, because it could be disputed whether it was "sufficiently established to have gained general acceptance in the particular field in which it belongs." However, because of the liberalization of the use of expert testimony in the courts, this kind of evidence based on syndromes and the like is increasingly being accepted by the courts as valid.

The result of this and similar cases is that over the years there has been a gradual lowering of standards for expert testimony, and more and more experts are being called into the courts to give testimony. For example, in criminal cases, psychologists and psychiatrists may be brought in for both sides, and the result is a battle of the experts. The outcome of the trial may depend on which expert the jury finds more credible and thus on the skills of cross-examination of the attorney charged with questioning the expert. In some recent cases, judges are even allowing jury members to ask questions (see Younger 1982).

One danger in the use of expert testimony in legal argumentation is that the expert witness is generally brought in to testify for one side and is usually paid for appearing. Therefore, the expert witness usually has an interest in supporting one side of the case. The use of expert witnesses in court has become so prevalent that there are even "professional expert witnesses" who derive much of their income from appearances in court. These practices present a real danger of bias in testimony (Younger 1982). Nevertheless, the trial process is an adversarial procedure explicitly designed to bring strong arguments to bear on one side of a case, assuming that the opposition will also bring strong arguments to support their side of the case. So expert testimony is not inherently biased, but it can unduly influence the opinion of a jury if the expert is highly credible and is not cross-examined effectively by the attorney and if that attorney fails to make a strong case for his or her own side with comparable experts. In some cases the courts will even allow an attorney to question the opposing side's expert witness specifically to establish bias.

Three bases for this allegation of bias have been cited by Graham (1977, 50): first, financial interest in the case at issue on grounds of remuneration for services; second, hope of continued employment; and third, prior testimony for the same attorney or same party. Thus, Graham argued that the percentage of a person's livelihood contingent on performance as an expert witness should be regarded as relevant evidence for establishing bias of an expert witness in court. This reveals an important connection between *ad hominem* and *ad verecundiam* arguments. The use of an *ad hominem* argument to impute bias on the part of an expert is one of the leading argumentation techniques for attacking an appeal to expert opinion.

That the subjectivity and dangers of expert testimony are recognized by the courts is evident in the arguments they allow in challenging this testimony. The lesson, then, is that any expert's pronouncement should not be treated as immune from revision or interpretation.

In order to get an accurate and useful opinion from an expert, it is necessary to have extensive discussions and explanations of the basis of the argumentation. Thus, while expert opinion is useful in argumentation, it has its limitations and, by its nature, is a kind of evidence that should be critically challenged. Experts are often wrong, and what they say should always be subject to careful interpretation for they are variably proficient at communicating their advice in a form that is clear and useful.

How does one differentiate between an expert and a layman? In general, a layman may be defined as someone who has only an average level of skill or knowledge in a particular field. A novice in a field is someone who has some training or experience but is still at the learning stage. The expert is a person who has sufficient knowledge and experience to have mastered the advanced skills of a particular domain of knowledge or experience. Experts not only have special skills, they are also proficient in their actions and have special ways of applying what they know to tasks in their area of expertise. Experts are good not only at recognizing problems in their area but also at recognizing whether those problems are solvable and at solving them when they are. Thus an expert is not only someone who knows something but someone who knows when he does not know something.

The Function of Presumption in Dialogue

A presumption is a kind of assumption advanced by one party in a dialogue, but in order for the presumption to function properly in the dialogue, the respondent to whom it is directed must also be involved. Thus, the putting forward of a presumption affects both parties in a dialogue—the proponent and the respondent. There are obligations on both parties to behave in particular ways, and the commitments of both parties in a dialogue are affected by the advancing of a presumption.[5]

A presumption is a move or speech act in a dialogue that is stronger than a pure assumption, in respect to the implications it has for commitment in dialogue, but weaker than an assertion. An assertion carries with it normally a burden of proof that falls upon the proponent of the assertion once he has brought the assertion forward in a dialogue. If challenged by the respondent, the proponent must either bring forward evidence to support his assertion or give it up, that is, retract it. With a pure assumption (supposition) this element of burden of proof is not present.

With a presumption, however, there is an aspect of burden of proof, like the assertion, except that the obligations or roles of the proponent and the respondent are reversed. A presumption is a commitment request put to a respondent by a proponent in dialogue, but it is up to the respondent to reject the presumption. Otherwise, it will be taken for granted in the subsequent dialogue that the respondent accepts the presumption as holding.

Presumption is easily confused with presupposition because both notions

5. See Rescher 1977 and Ullman-Margalit 1983, who put forward views of presumption that support the interpretation given here.

have one participant in a dialogue bringing forward a proposition for the other participant to accept for the sake of argument, without the first participant explicitly arguing for that proposition right at that point in the dialogue. For example, if I ask you, "Why did you steal Ed's bicycle?" my question has the presupposition that you stole Ed's bicycle. But this same proposition is also a presumption, in that if you give any direct answer to the question, or otherwise respond by not explicitly rebutting this proposition, it will be taken for granted that you are conceding it. So presumption and presupposition often refer to the same things in argumentation.

But presumption and presupposition have different functions in argumentation. In presupposition, the prefix "pre" refers to the sequence of moves in the dialogue just prior to the move in question. In presupposition, a speech act depends on or presupposes this prior sequence. Presupposition refers to past events in the sequence of dialogue; presumption is directed to the future sequence of moves in a dialogue: if the respondent fails to deny the proposition in question, his or her commitment to it will carry forward automatically over that future sequence (up to a point where either the dialogue ends or both parties agree to give up the presumption). In Lewis and Short 1969 (1433) the meaning of *praesumptio* is "taking beforehand, a using or enjoying in advance" or "a taking up and answering in advance, an anticipation of possible or suspected objections." Presumption always refers forward to the projected sequence of moves in a normative model of dialogue.

Presumption in dialogue enables argumentation to move forward in a revealing and useful way even if sufficient evidence is not available to enable either side to commit itself categorically to some particular premises. It enables one side to put arguments forward and the other side to hear and possibly criticize these arguments, even if both sides agree that the premises cannot definitely be proved or disproved at the present state of knowledge.

Presupposition enables each move in an argumentative exchange to be relatively simple and manageable by virtue of depending on previous moves in the dialogue and thus on previously determined matter. For example, I could ask you, "Why did you steal Ed's bicycle?" as an appropriate question in a dialogue if, at some prior move, I had secured your commitment to the proposition that you did steal Ed's bicycle. Once the one concession is fixed in place as a commitment, the asking of the second question can be built on that prior exchange. The second question is not required to be "iffy" or otherwise complex.

Such a notion of presupposition is clearly pragmatic. It refers to the prior sequence of a series of exchanges in dialogue. It is a dependency relationship composed of a sequence of speech acts in a dialogue.

A presumption is a proposition put forward by a proponent that shifts an obligation onto a respondent in dialogue. If the respondent wants to reject the proposition, he is obliged to present evidence to rebut it. Otherwise, the presumption is lodged in place as a commitment of both parties in the dialogue, until such time that evidence comes in or is brought forward that is sufficient to refute it. A presumption then could be described as a tentatively placed commitment in dialogue agreed to by both parties; if the respondent doesn't voice disagreement, it is taken for agreement.

Presumption can be defined as a speech act that contributes to the moving ahead of a dialogue through seven stages.

1. Argumentation is taking place in a given context of dialogue involving two participants called the proponent and the respondent, each of whom has a particular role to play.
2. At some particular point in the dialogue the proponent brings forward a proposition as an assumption that is useful for her argument, asking the respondent to adopt it as a provisional commitment.
3. The respondent has the choice of rejecting it, but if he does not, the proposition is immediately inserted into the commitment sets of both participants, subject to rebuttal.
4. If the respondent is now to successfully rebut the presumption, that is, avoid commitment to it, he must bring forward sufficient positive evidence or reasoning. Rebuttal now has a "cost."
5. Later on, at any subsequent point in the dialogue, if either party wishes to reject the presumption, he or she can do so by bringing forward evidence against it. If new evidence becomes available in the dialogue, the presumption can be rebutted or altered, according to this evidence.
6. The presumption stays in place during the dialogue until such time as it is rebutted by the agreement of both parties according to the requirements of 5. Once in place in the dialogue, it stays in place until retracted.
7. The presumption can be used as a premise, carrying the commitment of both parties, in the argumentation of either party during the course of the dialogue, as long as it has not been rebutted under clause 5.

Thus, a presumption secures provisional commitment for the sake of argument so that a dialogue can move toward its goal without the proponent of an argument having to meet impossible or impractical standards of proof, standards confounded by a lack of decisive or irrefutable evidence or by a lack of resources.

Presumption is a useful concept in argumentation for various reasons. If there is controversy about some problem like using nuclear energy or the ethics of abortion, where the propositions at issue are about values or personal preferences that cannot be decisively proved or refuted by factual evidence, it may be useful to base argumentation on presumptions that both parties can agree to. Otherwise, rational discussion could not get far, or would even be impossible. But it can still be useful and enlightening, in some cases, to see how the argumentation on both sides will go, or who has the strongest case, even though "the facts" are not all "in" yet on the issue.

There are six noteworthy bases for making presumptions that show the practical value of arguing on a basis of presumption. First, provisional assumptions are often useful to facilitate action that may be required in a situation where there is no time for lengthy inquiries. Second, presumptions are often based on routine ways of doing things that have been found successful in the past (for example, medical practice is often based on customary ways of doing things that have been found acceptable and safe in the past). Third, presumptions in argumentation are often based on expert opinion. Fourth, presumptions are often based on popularly accepted ways of doing things—customs and fashions. Fifth, presumptions often reflect conventional wisdom—assumptions or beliefs that are taken for granted by "common sense" where there has been no real reason brought forward to question them. Sixth, presumptions often take the form of principles of social cooperation and politeness that make smooth collaboration in social activities possible. Speech acts in dialogue are generally based on Gricean principles of cooperation; hence, the concept of presumption has an important place in the metatheory of argumentation in dialogue.

Speech Act Conditions for Presumption

Presumptive reasoning has a negative logic that takes the form of the *argumentum ad ignorantiam*. If there is no definite knowledge or hard evidence that a proposition is true, it can be presumed (tentatively) false; contrarily, if there is no evidence a proposition is false, it can be presumed true.

So conceived, presumption is a kind of speech act that is halfway between assertion and (mere) assumption. An assertion carries with it a burden of proof on the proponent. An assumption can go forward without proof. An assumption can even be contrary to known facts. And anyone in a dialogue is free at any time to reject an assumption without having to disprove it. Like an assumption, a presumption does not need to be backed up by evidence. But like an assertion, it does bear a relationship to evidence. However, that relationship is an oblique or negative one.

The key thing about presumption is that it reverses the burden of proof by switching the roles of the two participants in a dialogue. Normally, the burden of proof is on the proponent of a proposition asserted. However, in the case of a presumption, a burden of disproof falls onto the side of the respondent. Otherwise the presumption stays lodged in place in the dialogue.

In the following analysis of the pragmatic logic of presumption in dialogue, the point at which a presumption is brought forward for consideration is called "move x." The point where the presumption may be rebutted or given up is called "move y."

I. Preparatory Conditions
   A. A context of dialogue involves two participants, a proponent and a respondent.
   B. The dialogue provides a context within which a sequence of reasoning can go forward with a proposition A as a useful assumption in the sequence.
II. Placement Conditions
   A. At some point x in the sequence of dialogue, A is brought forward by the proponent, either as a proposition the respondent is asked explicitly to accept for the sake of argument, or as a nonexplicit assumption that is part of the proponent's sequence of reasoning.
   B. The respondent has an opportunity at x to reject A.
   C. If the respondent fails to reject A at x, then A becomes a commitment of both parties during the subsequent sequence of dialogue.
III. Retraction Conditions
   A. If, at some subsequent point y in the dialogue ($x < y$), any party

wants to rebut A as a presumption, then that party can do so, provided good reason for doing so can be given. Giving a good reason means showing that the circumstances of the particular case are exceptional or that new evidence has come in that falsifies the presumption.

B. Having accepted A at x, however, the respondent is obliged to let the presumption A stay in place during the dialogue for a time sufficient to allow the proponent to use it for his argumentation (unless a good reason for rebuttal under clause III. A. can be given).

IV. Burden Conditions

A. Generally, at point x, the burden of showing that A has some practical value in a sequence of argumentation is on the proponent.

B. Past point x in the dialogue, once A is in place as a working presumption (either explicitly or implicitly) the burden of proof falls to the respondent should he or she choose to rebut the presumption.

In a dialogue a presumption gives the argument some provisional basis for going ahead, even in the absence of firm premises. Once the presumption is lodged in place, the respondent is obliged, out of politeness, to leave it in place for a while, giving the proponent a fair chance to draw a conclusion using it as a premise. How firm a commitment is put into place depends in a given case on the type of dialogue and other global factors like the burden of proof, as well as local requirements defined by the type of argumentation scheme used at the local level.

But generally, the sequence of speech act conditions above shows how the negative logic of presumptive reasoning should work in a context of dialogue. This set of conditions provides a normative framework that enables one to evaluate presumptive reasoning, as used in context of dialogue. It therefore also provides the tool needed to evaluate certain tricky kinds of argumentation as fallacious or nonfallacious in a given case.

Cognitive Value of Presumptive Reasoning

What then is the value of presumptive reasoning if it does not result in conclusions that are true or known to be true? Is it a purely subjective and therefore suspicious kind of argumentation, a stimulating and colorful exercise, but one that has no really worthwhile cognitive value compared to the hard evidence gained as the result of scientific inquiry?

First, it should be noted that presumptive reasoning is monitored or checked by hard evidence, but in a way different from knowledge-based reasoning. Presumptive reasoning's relationship to hard evidence is indirect. It entails reaction in dialogue—dialectical attitudes and commitments, as revealed by the defense of a point of view. When it comes to evaluating presumptive reasoning, it is a question of seeing how it is used in argumentation in a context of dialogue, how criticism is dealt with in the sequence of speech acts that go back and forth in the argumentative exchanges. A position or point of view, when it comes into conflict with hard evidence, is not necessarily refuted. Instead, typically, a burden of proof is shifted against it—see Johnstone (1978).

A critic who deserves to be taken seriously should be able to back up a specific criticism with strongly reasoned evidence, carefully documented so that the arguer criticized cannot easily avoid, overturn, or repel the criticism. This is hard to do well, and cannot be done in many cases as easily as the superficial critic may think. A reasonable critic must have empathy and must ask how the one criticized might reply to a criticism. This requires an insight into the other arguer's position. What is most important to avoid is a dogmatic rejection of an opponent's argument without giving due respect to the other arguer's point of view or reasons.

The primary difficulty in evaluating criticisms is for the evaluator to be objective, meaning that bias and dogmatism must be avoided. *Bias* is the tendency to distort the strength or weakness of an argument based on the critic's inherent commitment of his or her own position on the issue.[6] *Dogmatism* is the unwillingness to examine or consider the position on the other side of an argument. The dogmatic arguer refuses to concede the real weight of argument on the opposing side, to see the other side at all. A critic must try to be neutral and fair, but there is an inherent tendency for any critic to choose sides and to succumb to bias or even dogmatism.

Fallacies occur where a presumption is pushed ahead too aggressively or too dogmatically by a proponent, without leaving room for the possibility of qualifications, corrections, or critical questions. This occurs, in many cases, where a weak presumption is treated as much stronger than it really is, perhaps even as a required presumption that cannot be challenged or

6. Blair 1988. Bias is more fully analyzed in Chapter 8.

rebutted unless the case is an exceptional one. Most presumptions are weak.

There are three types of presumptions in arguments: required, reasonable, and permissible presumptions. A *required presumption* is one that is brought forward by the proponent as a proposition that must be accepted in all subsequent developments of the line of dialogue, unless it can be clearly shown by the respondent that the case is exceptional. For example, for anyone who picks up an uninspected weapon on a firing range, it is a required presumption that the weapon be treated as loaded. The person in question must always act in accordance with that presumption. A *reasonable presumption* is one that is brought forward by the proponent as a proposition that is to be accepted in normal or reasonably expected developments of the line of dialogue. For example, if Bob's hat is on the desk, it may reasonably be presumed he is at work, given what is known about Bob's usual routines. A *permissible presumption* is one that is brought forward by the proponent as a proposition that is optionally acceptable for the respondent; that is, it is meant to be accepted in some, but not necessarily all, subsequent developments of the line of dialogue. The respondent can reject it simply by indicating disagreement concerning its application to a given case. A permissible presumption is an assumption one party asks another to adopt as a provisional commitment for the sake of allowing a dialogue to go forward.

Of course, all presumptions are inherently tentative and provisional, subject to rebuttal or correction. But it does not follow that presumptive reasoning in dialogue is inherently fallacious or deceptive, completely unreliable as a basis for prudent action or for arriving at a conclusion, for the fallacies function as negative boundary conditions for presumptive reasoning in dialogue. A controversial presumption can be subjected to the hard testing of argumentative dialogue with an able opponent. If it survives the test, it can be strengthened, and many of its faults or false implications can be exposed. Through this process of examination, important new insights can arise.

The most important benefit of reasoned dialogue of the critical discussion type is the insight each party gains into the other's position (empathy) and the self-knowledge gained through deeper insight into the reasoning behind one's own position, testing it out in free argumentation against a critical opponent. It is this process of articulation and clarification of one's own personal, deeply held, but often inchoately expressed or "dark" commitments that is the most important cognitive gain of an excellent critical discussion. This personal insight, or deeper understanding of an issue, is not itself hard evidence. But it can prepare the way for knowledge—and is even perhaps essential for this purpose—by clearing away the rubble of fallacies, misconceptions, dogmatic attitudes, and superficially attractive but untenable points of view that won't stand up to serious cross-examination in discussion.

Why concentrate on the negative: the pathology of argument, the errors, weak points, or criticisms? One reason is that we human reasoners are, in many cases, not as reasonable or wise as we might like to think, especially on difficult topics like politics, religion, ethics, and national affairs, where it is only too easy to be blinded by personal self-interest, to give in to emotion, and to lose critical perspective. Personal attack, emotional diversion, and other temptations are more often characteristic of human reasoning than correct and coolly impartial arguments. No doubt this is simply because real learning is most often a process of trial and error. And our most impassioned and cherished convictions may be based more on personal familiarity, in many instances, than on analysis of reasons for and against. But another point to be made is that good criticism need not exclusively or ultimately be concerned with errors and bad arguments. A weak argument can be strengthened through criticism, and thereby often becomes a stronger argument. So curiously enough, the critical perspective often turns out to reveal moves in argument that are justified and correct.

Going through the various fallacies, it will be found that criteria can be given in the form of argumentation schemes to help differentiate between the arguments justifiably open to criticism and those that can be defended against criticism. If a critic claims that an argument is weak in this or that respect, it follows that a certain burden of proof rests with the critic to give evidence from the *corpus* of argument to show which specific guideline of reasonable argument has been violated in this particular case. The goal of informal logic should be to show how the reasonable critic of an argument can fulfill this burden of proof by giving reasons for his criticism, based on guidelines for reasonable argumentation in a normative model of dialogue.

3

ARGUMENTUM AD POPULUM

It is the contention of this chapter that not only is the appeal to popular sentiment or opinion of the type associated with the traditional *argumentum ad populum* a nonfallacious kind of argumentation in some contexts of dialogue, it is a legitimate technique and can be an important part of constructing a correct and successful argument. This will be shown to be especially true in certain contexts of dialogue, particularly those of political argumentation in a democratic system.

The point of view will be pragma-dialectical. That is, arguments will be judged as correct or incorrect, successful or unsuccessful, as used in a context of dialogue to support the relevant goals. The set of procedural rules for a particular type of dialogue is taken to provide a normative model that can be applied to a given text of discourse in order to judge an argument in that discourse as correct or incorrect.

A fallacy is taken to be incorrect argumentation, but not just any weak or lapsed argument. A fallacy is a serious, underlying, systematic error in the reasoning (the chain of logical inferences) in an argument or a sophistical tactic to get the best of a partner in dialogue. A fallacy, then, is a technique of argumentation that has been used wrongly (abused) in such a way that it goes strongly against the legitimate goals of a dialogue.

The chapter begins with a critical examination of five major reasons given by the logic textbooks for condemning ad populum argumentation as fallacious. The first is that the ad populum argument is an appeal to emotion, the mass enthusiasms or popular sentiments of the crowd. The second is that the ad populum is a subjective argument that directs an appeal at the prejudices or sentiments of a particular audience the argument is designed to persuade. The third is that ad populum argumentation is a failure of relevance insofar as it tries to evade, cover up, or substitute for a failure to bring forward good evidence or reasons by appealing to the enthusiasms of the multitude. The fourth is that the ad populum assumes a direct relationship between an argument's validity and its popularity. The fifth is that the use of the ad populum argument involves an illicit shift from the critical discussion of an issue to a self-interested bargaining or negotiation dialogue that appeals to members of a special interest group while excluding all those who do not belong to that group from the dialogue.

After examining each of these reasons, it is concluded that none of them individually is sufficient to explain why the argumentum ad populum is a fallacy.

Next, it will be shown how argumentum ad populum can be a reasonable, nonfallacious argument, in some cases. And then finally, it will be shown that the ad populum argument is used fallaciously where it is employed as an overly aggressive tactic that goes against the legitimate goals of a dialogue.

Appeals to Emotion

What seems essential to most accounts of the ad populum fallacy given in the textbooks is that the fallacy involves an appeal to emotion. The idea is that this argumentation that appeals to popular sentiments is wrong precisely because it is an emotional appeal as opposed to an appeal to logical reasoning. The one is seen as incompatible with the other. For example, according to Copi (1986, 96), the argumentum ad populum is "the attempt to win popular assent to a conclusion by arousing emotions and enthusiasms of the multitude" instead of appealing to the facts in an argument. What is fallacious is the attempt to utilize the mass emotional appeal in place of logical argumentation.

There are two problems with this characterization of the *ad populum* fallacy. One is that it would be difficult to define the fallacy in terms of attempts to arouse mass enthusiasms, because this aspect is based partly on the intent of the proponent and partly on the emotional response of the audience. Such an approach makes the fallacy a psychological and empirical matter for which it would be difficult to devise logical decision procedures or rational guidelines to distinguish between the fallacious and nonfallacious cases. The second problem is that this approach rests on the presumption that appeals to popular emotions are in a separate category from logical reasoning. The suggestion is that appealing to the emotions of the respondent is never a good way to argue.

According to a rationalist viewpoint often expounded in traditional philosophy, rational argument should exclude emotional considerations as irrational "passions" that "fog the mind" and bias argument instead of basing argument on impartial reason. According to this viewpoint the use of emotional appeals in argumentation is always suspicious, if not outright fallacious. In opposition to this severe view, others have argued that emotions have some legitimate role to play in argument, especially in ethical or political deliberation, in spheres of argumentation where personal loyalties and commitments are involved in a decision, and in controversial issues where hard knowledge of the facts is not decisive in leading to a single conclusion. Callahan (1988), for example, has argued that there should be a mutual interaction of thinking and feeling in ethical decision making, whereby reason and emotion tutor and monitor each other.

This ambivalence about the role of emotion in argument is reflected in the work of Aristotle. In the first chapter of Book I of the *Rhetoric* (1354a) Aristotle complained about the practices of rhetorical treatises that put too much emphasis on appeals to emotion, which "warp" or "pervert" a judge by personal appeals that have nothing to do with the essential facts of a case. However, in the second chapter of Book I Aristotle put *pathos* on the same level with *logos* and *ethos* as acceptable means of rhetorical argumentation (as Brinton [1988, 207] pointed out), and he took *pathos* seriously as a means of persuasion throughout the rest of the *Rhetoric*.

But how could appeals to emotion play a legitimate part in reasonable arguments? One need only break free from the deductive paradigm of valid argument in scientific inquiry as the sole legitimate context for successful argument to find places for emotional appeals in argumentation. In practical reasoning, action is required of an agent who typically has to start from presumptions in a situation where time does not allow for an exhaustive inquiry to establish indisputable knowledge. The agent may also have to act on presumptions, in deciding the best or most efficient way to do something. And if the agent is not an expert on doing this sort of thing, he or she may have to rely on the advice of others. In some cases the agent may even be well advised to rely on customary or usual ways of doing something, as expressed in popular opinions or sentiments. In such cases "gut feeling," or an appeal to emotion, may be a reasonable alternative to, say, flipping a coin.

Acting on presumptions is a reasonable kind of argumentation or deliberation if carried out according to standards of burden of proof appropriate for the context of dialogue. Even if an emotional appeal is best seen as somewhat unreliable or unstable, not a good substitute for harder evidence when it is available, such an appeal need not be excluded altogether from the realm of reasonable argumentation.

In a legal, political, or ethical discussion where the subject concerns the character or personal morality of the participants, appeal to emotion may not only be relevant to the issue, it may be absolutely required in order to respond effectively and convincingly to an allegation.

Although many of the moral philosophers, most notably the Stoics, have been against anger, seeing it as a morally improper response to any situation, Aristotle distinguished between justified and unjustified anger. In the *Nicomachean Ethics* (1106b), Aristotle wrote that anger, like other emotions, may be felt too much or too little, and that the middle way, or "mean," appropriate for the situation is the virtuous response. Thus, for Aristotle, not feeling angry in response to an unmerited insult indicates a defect of character. This point of view led Brinton (1988b, 81) to postulate a kind of argument called the *argumentum ad indignationem,* a pathotic argumentation that gives legitimate vent to anger when the feelings expressed and the degree of intensity with which they are expressed are appropriate.

Whether or not the *argumentum ad indignationem* is a distinctive type of argument that should be added to other arguments *ad,* it certainly has application as the basis of the principle underlying shifts in burden of proof in the tactics of argumentation. As Whately ([1846] 1963, 113) observed, failure to react strongly enough, to defy your accuser to supply proof, when an unsupported accusation is brought against you in argument, is a serious tactical error that can make you appear to be guilty of the offense alleged. You must react strongly and shift the burden of proof back onto your opponent, or you may come out of the exchange poorly in the eyes of your audience.

In the classic FED (Federal Election Debate) case presented on pages 83–84, one speaker in a political debate charged that the other speaker had "sold out" the country. Indignant at this personal attack, the other speaker launched into an emotional *ad populum* appeal expressing his personal loyalty as a patriotic citizen. By the current standards of the logic textbooks, it would not be out of place to cite exactly this sort of case as an instance of the fallacious *ad populum* appeal. But viewed in its context of dialogue—as a response to a personal attack, an accusation of disloyalty in the course of a political debate where character and leadership are legitimate issues in the dialogue—the emotional appeal does not seem so out of place or irrelevant.

Any good theory of argument must recognize that emotional appeals and responses need to be balanced against cognitive considerations. But on the other hand, in most of the important personal decisions in life, sensitivity to emotions and feelings of one's own, as well as of others, is extremely important. Any theory of argument that rules such appeals out of court altogether must be a very limited theory, inapplicable to many, significant everyday arguments. Especially in presumptive reasoning, which is so common, for example, in controversial political debates concerning choice of policy or candidate, gut feeling and emotional reaction, always subject to critical reflection, have an important place. To declare that any conclusion resting even partly on an appeal to emotion must be fallacious is an implausible and even misguided approach.

Audience-Directed Argumentation

One common explanation of what is wrong or fallacious with the use of *ad populum* arguments is that this type of argumentation is directed to a specific group of respondents instead of being a universally valid argument binding on any rational respondent. The objection is that it is a partisan appeal, meant to persuade a specific audience, and that it is successful to the extent that it succeeds in persuading this specific audience, even if it panders to the prejudices or idiosyncratic position of this audience. With an *ad populum* argument, what matters is not whether the premises are true, or based on good, objective evidence, but whether the audience to whom the argument is directed accepts these premises enthusiastically (Walton 1980, 267).

According to this account of *ad populum* arguments, what is wrong is that they throw concern for truth aside, in favor of any premises (even false ones) that the target audience is prepared to accept. Argumentation goes over to a person-relative salesmanship that has no regard for arriving at truth or knowledge by appealing to objective evidence.

Much the same account could be given about what is fallacious with *ad hominem* arguments or other person-relative kinds of arguments associated with the traditional informal fallacies.

What is open to criticism about this account is the presumption that the only legitimate premises of a sound argument are those that are true, or known to be true. What is overlooked is the dialectical argument involving two participants in a dialogue exchange where one party utilizes premises that are commitments of the other party in order to reasonably convince that party to accept a conclusion that is in question or subject to doubt. For such an argument to be successful in fulfilling an appropriate burden of proof, it is not necessary that the premises be incontrovertibly true, or known to be true. Such premises can be presumptions that are commitments of one party, even though they are controversial and cannot, at the present state of knowledge, be proved by appeal to facts, or evidence that has been established as knowledge.

In the critical discussion, a type of persuasion dialogue, the goal is for one party to persuade the other to accept one's conclusion, using premises accepted by the other party, in order to resolve a conflict of opinions (see Van Eemeren and Grootendorst 1984). Argumentation by each party in a critical discussion is directed toward the other party as respondent and is based on the commitments of that specific respondent. But it does not follow that all arguments in a critical discussion are fallacious *ad populum* arguments. Far from it. What is shown is that argumentation directed to a specific respondent is not necessarily fallacious. It can be a reasonable type of argument within a given context of dialogue.

Of course, an argument that is reasonable in one context of dialogue may be used fallaciously in a different context of dialogue. In the type of dialogue called the *inquiry*, the goal is to establish premises that are known to be true, by an orderly process of investigation, and then use these premises to move forward and prove a conclusion. The goal of the inquiry is to prove a conclusion beyond reasonable doubt (according to a standard of burden of proof appropriate for the type of inquiry involved), or alternatively, to show that this conclusion cannot be proved at the present state of the advancement of knowledge.

Aristotle's *demonstration*, a type of argumentation identified at the beginning of the *Prior Analytics*, is very similar (or perhaps identical to) what is called the inquiry above (64 b 33). Aristotle also distinguished between the demonstration and the type of argumentation he called dialectical. According to Aristotle, a demonstrative premise is one that is laid down as a first principle, whereas a dialectical premise, usually an opinion of the many or the wise, is one that one's opponent in dialogue is prepared to admit as a basis for argument (*Topics* 101 a 37).

Ad populum argumentation would generally be out of place and rightly regarded as inappropriate or fallacious in an inquiry. But is it so obvious that it would always be inappropriate or erroneous in a critical discussion? Not at all. In fact, if one's opponent in a critical discussion consists of a group of people or a mass audience (for example, in a public speech or debate format), then arguing from commitments of this group is precisely the kind of argumentation that is correct and appropriate. If the goal is to convince this group by reasoned persuasion, then the commitments of the group must be taken as the starting points from which to select the arguments' premises, if one's argumentation is to be successful.

What is rightly brought out by the specific-audience theory is that *ad populum* argumentation can be fallacious in those instances where it is used improperly or in an inappropriate context of dialogue. What is wrongly inferred from this theory is that it shows exactly what is wrong with an *ad populum* argument when it is wrong or fallacious. Also wrongly inferred is the conclusion that *ad populum* argumentation is inherently wrong in all instances where it is used.

One might still object, however. How can dialectical argumentation, which is person-relative, be reasonable or correct if it is based only on presumption and burden of proof? Doesn't valid argumentation require a "universal audience"?

In order for adjustments of burden of proof in an argument to be "reasoned," the two-sided dialogue must have procedural rules that are "reasonable" or that somehow represent, or are related to, rules of logic. The concept of reasoned dialogue required must have normative models containing the kinds of rules described in Walton (1989, chapter 8): locution rules, dialogue rules, commitment rules, and win-loss rules. But are these normative models enough to secure the distinction between persuasive rhetoric and reasoned conviction in a given case?

Perelman (1982, 17f.) went to the heart of the matter by making the distinction between persuasive discourse addressed to a few, that is, a specific, target audience, and discourse that purports to convincing generally, that is, valid for everyone. Perelman called the first kind of discourse *persuasion*, as opposed to the second kind of discourse, called *conviction* (or *convincing discourse*). Through the second kind of discourse, Perelman (1982, 18) introduced the idea of the *universal audience*, meaning the class of all reasonable respondents to whom a convincing argument is directed. According to Perelman: "A convincing discourse is one whose premises are universalizable, that is, acceptable in principle to all the members of the universal audience." This distinction between convincing and persuading discourse seems fundamental to understanding the *ad populum* argument.

However, there are important questions about whether the "convincing" type of discourse based on the universal audience is necessary to validate the critical discussion. It could be that dialogue with the goal of convincing a universal and perfectly rational respondent is an ambitious ideal that is appropriate for the inquiry but not for critical discussion. Some might claim that scientific reasoning is an instance of convincing dialogue directed to a universal audience. However, it could well be that the kind of knowledge-based deductive and inductive reasoning so often held to be characteristic of scientific knowledge, although appropriate for the inquiry, is too high a requirement to be suitable for argumentation in a critical discussion.

Others might claim that philosophical argumentation, for example, the Platonic dialogues and other classical philosophical texts, is an instance of "convincing" dialogue directed to a universal audience, but there are grounds for having reservations about these claims as well. It may be that all persuasion dialogue argumentation in natural language presupposes a narrative and historical context of background presumptions and unstated premises that can be understood and accepted only by a specific audience or readership, even if that audience cuts across historical epochs and cultures.

Thus, it would not seem to be necessary to require a universal audience for argumentation in persuasion dialogue. Knowledge-based interactive reasoning would only be necessary and useful if all the participants in the dialogue shared the same (universal) knowledge base. Hence the universal audience is a kind of fiction, as far as dialectical argument is concerned, never met with or required in the practices of persuasion dialogue.

The universal audience could possibly be an appropriate model for inquiry in, for example, the natural sciences, where Locke wrote of reasoning based on probability and the "natural light of evidence" from nature. But persuasion dialogue yields insight (or reasoned commitment) based on different positions for different participants, and this concept of interactive reasoning is the key to making sense of the *argumentum ad hominem, argumentum ad verecundiam,* and *argumentum ad ignorantiam.* The best that persuasion dialogue need aspire to is to provide conclusions that are reasonable presumptions, based on interactive argumentation schemes that are inherently open to critical questioning in subsequent dialogue. Such conclusions can be justified as reasonable guides to action, even in situations where knowledge is incomplete, for three reasons: (1) they are based on normative models containing procedural rules of reasoned dialogue at the global level; (2) they are based on the use of argumentation schemes at the local level; and (3) they elicit knowledge through the maieutic function of reasoned dialogue, which clarifies and articulates the deeply held but "dark" or hidden commitments of the participants.

So conceived, the goal of critical discussion, unlike inquiry, is not to prove conclusions based on knowledge or established, verified facts. It has the more modest goal of exposing errors, fallacies, dogmatism, and superficial beliefs by putting them to the test of critical questioning in dialogue with an able opponent. In this framework, it can by no means be taken for granted that *ad populum* arguments are always or inherently fallacious.

Relevance

Most textbooks place importance on failure of relevance as a factor in explaining the *argumentum ad populum* as a fallacy. Castell (1935, 24) states that the "root objection" to the *ad populum* is that it "substitutes an appeal to popular sentiment for an appeal to relevant facts." Copi (1986, 96) classifies *ad populum* under fallacies of relevance and defines it as a fallacy on grounds of its being an appeal to the "enthusiasms of the multitude" as opposed to an appeal to "relevant facts." Engel (1976, 121) after quoting Mark Antony's speech at length, describes it as a "cunning introduction of irrelevancies" that commits the fallacy of mob appeal.

These treatments of the *ad populum*, despite the plausibility of categorizing the failure as a lack of relevance, do not clearly define relevance. Failure of relevance has become a kind of "wastebasket category"—if a fallacy can't be diagnosed as an error by any other means, it is an easy dodge to classify it as a failure of relevance. But what kind of a failure is that? Presumably most, or perhaps all, fallacies can be categorized failures of "relevance" in some broad sense. But that doesn't explain what, in particular, makes an *ad populum* argument necessarily fallacious.

The other problem with this approach to *ad populum*, the textbook explanation of this fallacy, is that it does seem that appeal to popular sentiment or emotions can often be "relevant" in many contexts of dialogue, like political speeches or funeral orations; it therefore remains to determine when such an appeal is relevant and when it is irrelevant.

However, as so many of the textbooks allege, the *ad populum* is surely a fallacy of relevance. But saying that the *ad populum* is a fallacy because it is a failure of relevance is not the whole story of the *ad populum* as a distinctive type of fallacy, for there are many other fallacies, like the *ad hominem*, *ad misericordiam*, and so forth, that are also failures of relevance. What is distinctive about the *ad populum* as an individual fallacy of appeal to popular sentiment? And what is relevance? And when is irrelevance so bad that it become fallacious? The linking of *ad populum* to irrelevance in argumentation is not a bad first step, but it does not (by itself) provide an analysis of what is fallacious about *ad populum* arguments.

Current research suggests that a pragma-dialectical concept of relevance can be defined by presuming that each type of reasoned dialogue has a goal, and argumentation in that context of dialogue is relevant to the extent that it contributes to this goal. In a critical discussion, for example, the goal is to resolve a conflict of opinion between the two participants through question-reply argumentation according to the procedural rules of dialogue for a critical discussion. This means that the participants each have a goal of proving or questioning a particular proposition (or point of view). To do this, they ask each other questions, taking turns, and advance arguments, using accepted argumentation schemes, in order to fulfill their burden of proof in the discussion. Deviations from these goals and obligations of proof will be considered *irrelevant* moves in the dialogue.

For example, suppose a participant in a critical discussion on a housing shortage has the burden of proving to the satisfaction of her audience that a particular solution to the shortage is best. Suppose further that in the middle of one of her speeches on the housing problem, she launches into a quite general, but very emotional, argument to the effect that all people deserve decent housing. This tub-thumping declaration smacks of popular piety that nobody would disagree with and that doesn't require proof at all in this context. Why is it in her speech? The suspicion is that it has no real function in fulfilling the requirements of burden of proof for this speaker. It's just there because it is a rousing appeal to popular sentiments, values, or interests that will generate mass enthusiasm in favor of the speaker.

In this type of case then, where the speaker's emotional interlude is not relevant to the proper fulfillment of her real obligation of proof in the dialogue, she has committed the *ad populum* fallacy.

This approach to relevance as a pragma-dialectical category of argumentation evaluation can lead one step further toward an understanding of the *ad populum* argument as a technique that is subject to misuse in some cases, in relation to the goals and requirements of a dialogue type. But until the general problem of relevance is solved in the field of argumentation, many questions will remain open.

In some cases, an argument seems irrelevant because it fails to contribute to legitimate goals in a dialogue and is therefore a weak or virtually useless argument for the purpose it is supposed to fulfill. In other cases, a failure of relevance is so bad that the move made in an argument seems to go out of the proper context of dialogue altogether. In the latter case, there is a dialectical shift from one context of dialogue to another. A tub-thumping declaration that all people deserve decent housing might be "relevant" in some sense—it may be, for example, relevant to the speaker's own need to affirm her gut feelings and personal emotions of the moment as a "woman of the people." But it may not be dialectically relevant to the fulfillment of this speaker's obligation to show that her solution to this particular housing problem is better than the proposed solutions of her competitors.

To deal with the *ad populum* as a fallacy in this type of case it is necessary that "relevance" be understood as *normative relevance*, meaning the relation of a speech act to goals of a discussion that a speaker is supposed to be engaged in. Such a meaning makes relevance relative to the context of a discussion. An *ad populum* appeal may be relevant in one context of dialogue, irrelevant in another. Appealing to relevance then does not imply that *ad populum* argumentation is fallacious in all cases.

The Fallacy of Popularity

Johnson and Blair (1983, 157) describe a form of inference they call the *fallacy of popularity*: everyone believes A, therefore A is true. The flipside or negative variant is the inference: nobody believes A, therefore A is false. According to Johnson and Blair these two types of inference are so outrageous as moves in argumentation that they rarely occur in this "blatant formulation," most often appearing in a format where one has to dig beneath the surface of the argument to uncover them.

But are these two types of inference fallacious? Johnson and Blair draw back from making this general claim, writing, "We are not about to propose that the popular acceptance of a belief is never any reason for thinking it is true" (158). They cite the following case. You find that everyone in a community you are visiting believes that the fish in a local lake are contaminated. Johnson and Blair concede that this might be "some reason" for you to believe that the fish are contaminated.

How can it be then that the two forms of the inference from popularity are sometimes fallacious and sometimes not? And if they are not fallacious in some instances, how can it be proper to label them as the *fallacy* of popularity?

The beginning of an answer to these vexing questions is perhaps to be sought in the following considerations. If, for example, the inference from popularity is taken to be a deductively valid inference that establishes the truth of its conclusion, it is easy to see why the inference is fallacious: mere belief can never—in most cases, at any rate—prove the truth (or falsehood) of a proposition. But if these two inferences are taken as defeasible inferences that can provide a plausible, but provisional, basis for prudent action in the absence of hard evidence, they could be quite reasonable in many cases.

Johnson and Blair's contaminated fish example is a good case in point. If you are a visitor to this local community, and you do not have direct access to any results of a scientific or official inquiry into the possible contamination of this lake, prudent action would be not to eat fish known to be from that lake. For reasons of personal safety, it is wise, perhaps, to place the burden of proof against the side of assuming it is safe to eat these fish. The cost of eating them could be high, while the cost of avoiding eating them is presumably low. Such a conclusion would be a defeasible one, however, based on popular say-so as a source for presumption that provides a guide for prudent action. It does not follow, in any stronger sense, that you "must believe" the fish are contaminated, as though you had been shown conclusive or incontrovertible evidence.

So the two inferences from popular opinion cited by Johnson and Blair may not be fallacies or fallacious inferences *per se*. They seem to become fallacious when exaggerated, overestimated, or put in a context of argument where they are used inappropriately. Properly seen or used as defeasible inferences that can form a tentative basis for prudent action in cases where knowledge is lacking, they can stand behind reasonable arguments. What makes their use fallacious in a particular case is a hardening into dogmatism, taking them as iron-clad arguments for the truth of a conclusion instead of defeasible, presumptive inferences. Instead of labelling the pair of inferences identified by Johnson and Blair as fallacious *per se*, calling them the "fallacy of popularity," one shall call them the *argument* (or *inference*) *from popularity*, a kind of argument that can be used fallaciously in some cases.

But there is more to Johnson and Blair's contaminated fish case than meets the eye. The local people's universal belief that these fish are contaminated carries weight in an argument pertaining to the safety of eating these fish because presumably these people are familiar with this local area and have a strong interest themselves in knowing about the safety of eating these fish. Therefore, if they believe the fish are contaminated, it may be wise to presume there is something in it. It is not that these local people are experts, but if experts were consulted, as would be likely if the contamination of the fish was an issue, then these people would be likely to know what the experts had reported.

Thus the contaminated fish argument is not just a simple argument from popularity, "Everybody believes it, therefore it is true." Additional premises concerning the backing of this belief seem to be acting as background presumptions in this particular case.

In fact, it often seems that the argument from popularity is not put forward in the bare-bones schema outlined by Johnson and Blair. More typically, it seems to have this argumentation scheme.

P1 Everybody (or everyone in some group) accepts A as a true proposition.

P2 These people are in a position to know that A is true, or at any rate, presumably have some reason for accepting A.

Therefore, A may be accepted as true.

The negative counterpart to this argumentation scheme for the argument from popular opinion starts from the premise "Nobody (or nobody in some group) accepts A as a true proposition." But there is also a stronger negative counterpart argumentation scheme, taking as the first premise "Everybody (or everybody in some group) accepts A as a false proposition." The weaker negative argument is a species of *argumentum ad ignorantiam*. In both negative arguments, the word "true" in P2 is replaced by "false."

Also, stronger or weaker variants of the negative arguments from popularity can be devised by distinguishing between the weaker form of the conclusion, "A may not be accepted as true," and the stronger form, "A may be accepted as false."

In the argument from popular opinion as used in everyday argumentation P2 often functions as a background presumption that may not be explicitly stated. An interesting case in point is the common consent argument for the existence of God. The historical pros and cons of this argument have been chronicled by Edwards (1967) in a survey, briefly summarized below.

Case 3.1

> This argument has rarely been stated in the form of appeal to the universality of belief in God, given the obvious objection that there is no reason why all mankind might not be wrong—truth does not follow simply from common consent. Another premise is added, which can take one of two forms, stating either that the belief is instinctive, or that the believers used reason in arriving at this conclusion. Some philosophers maintained, however, that neither premise is a good reason for accepting the conclusion that God exists. To support the second line of argument, some philosophers contended that the believers tend to be the more educated people.

This historical controversy is interesting for the student of *ad populum* argumentation because it suggests that the argument from popularity is usually backed up in practice by some additional premise that asserts the group has some particular reason for accepting the proposition in question.

It is also interesting to see the link with the *argumentum ad verecundiam* and related testimony-based types of appeal, where the group whose opinion is cited is said or presumed to have special access to or knowledge of the matter.

Although the argument from popularity looks like a fallacy when abstracted from a context of use, in fact, when expressed more fully through the argumentation schemes above, it quite often appears to be a reasonable kind of argumentation. Considered as a way of shifting a weight of presumption rather than as a deductively valid argument that goes simply from universal belief to the truth of a proposition, the argument from popularity no longer seems like a fallacy. Or at any rate, it seems misleading and simplistic to say that the argument from popularity is always fallacious—particularly when used in a circumspect and cautious way to shift a burden of presumption on an issue of dispute where overriding or decisive knowledge is lacking.

Also, it doesn't seem quite right to simply equate the argument from popularity and the *argumentum ad populum*. The former certainly seems to be one important ingredient in the latter, but it is not the whole. As well as an appeal to popular opinion or widely accepted belief, the *argumentum ad populum*, at least as traditionally conceived in the logic textbooks, also involves an element of emotional appeal to "mass enthusiasms," popular sentiments, or folksy pieties: a political campaigner "talks down" to his audience by emphasizing his alleged working-class origins, or his fondness for pork rinds, for example.

The Dialectical Shift Theory

In modern democratic politics, financial interests are always involved beneath the surface. Therefore, there is always a certain degree of cynicism or worry about any kind of political debate—what appears on the surface to be a critical discussion of the issues may mask self-interested bargaining or negotiation dialogue. Concessions to special interest groups or constituencies are a natural part of any political process in which a politician needs to court the votes of the majority to hold office.

Inevitably, voters don't understand all the specifics of every issue, nor can they anticipate the nature of future issues that will arise during the term of office of an incumbent. Therefore, they want to elect someone who seems likely to represent their interests on any issue—someone who has the same likes and dislikes as they have, someone who comes from the same background, someone who can be trusted to respond in the same way they would respond. In a nutshell, a voter wants to elect someone who has that voter's real interests "at heart," someone who can be counted on to support the interests of the audience to whom he speaks in order to get their vote. When a candidate goes into a political speech with such emotional catchphrases as "man of the people" or "I'm just a country boy" or "My grandaddy was a salt miner," this mode of speech may have a certain legitimacy in the context of dialogue. This political speaker is trying to affirm his solidarity with the people he takes to be the broad base of his supporters by saying, in effect, "I am really one of you," or "You can count on me." Nothing is inherently wrong with this kind of appeal, given the context of dialogue for political debating and campaigning in a democratic system.

On the other hand, this type of *ad populum* appeal is, by its very nature as part of a political speech, open to skepticism and suspicion of deceit or trickery. For, as noted above, it is based on a kind of self-interested bargaining dialogue that lies under the surface of critical discussion of the issues. Those interest groups who are not part of the bargain, and feel left out of the appeal are naturally going to be quite suspicious of it, and are going to feel such a speech is wrong, unacceptable, or biased. Even those who have nothing to gain or lose are going to be skeptical about the objectivity or fairness of this kind of speechmaking. Hence this type of *ad populum* argumentation in political debating is inherently open to cynicism and critical reactions.

Taking this approach, what is incorrect or fallacious about such an *ad populum* appeal is that there has been an illicit shift or mixing of two types of dialogue. On the surface the political debate appears to be a critical discussion of the issues. Under the surface it is a negotiation. Therefore, a political speaker who uses the *ad populum* argument has covertly and unfairly shifted from one type of dialogue to another. The criticism here is similar to that of the bias or "poisoning the well" type of *ad hominem* argument. This speaker is biased. He pretends to be fairly and impartially looking at both sides of the issues. But covertly, he is pushing for his own interests and for those who have a financial stake in his policies. His rhetoric is merely a deception and that's why his *ad populum* argumentation is fallacious.

This explanation reveals something important about the nature of *ad populum* argumentation in political debating, but it does not show that all *ad populum* appeals in political debates and speeches are inherently fallacious. The fallaciousness of the appeal depends on the context of dialogue of political debate.

What is political debate supposed to be? What kind of dialogue is it supposed to be? If it is supposed to be purely a critical discussion of an issue, or the issues, than the *ad populum* argument (on the explanation of it as a type of argument given above) is always fallacious, erroneous, or inappropriate in a political debate. If a dialogue is supposed to be a critical discussion—if that is the proper normative model of dialogue for evaluating argumentation—then negotiation dialogue may rightly be deemed inappropriate, incorrect, or even fallacious in relation to the goals and norms of correct argument in that model.

But is political debate a pure critical discussion? It seems much more characteristic of political debate in a democratic system that it is properly a mixture of several kinds of dialogue. It is supposed to be a free marketplace for the arguing out of policies and ideas, a "bear pit" where proposals can be attacked and criticized in public. On the other hand, it is not supposed to be a completely adversarial free-for-all where anything goes and all arguments are equally good. Critical standards of correct or fallacious argumentation are an important part of what makes political dialogue useful and instructive. But negotiation, or self-interested bargaining, is tolerable to some extent and is to be expected in any realistic political debate on a live issue.

Accordingly, it is proposed here that political debate in a democratic system of government is a mixed type of dialogue, composed of five types of dialogue that all have a legitimate place.

1. Information-seeking Dialogue. The purpose is for opposition members to ask for information on relevant issues of concern.
2. Action-producing Dialogue. The goal is to facilitate or press for action on urgent issues.
3. Eristic (Contentious) Dialogue. Question periods and election debates should allow for adversarial (partisan) exchanges.
4. Critical Discussion. Questioning of assumptions, and other clarifications and rebuttals, should be regarded as legitimate where they are appropriate.
5. Negotiation. Underlying political debates are very real conflicts of interests, in many cases, which may be a significant factor in partisan bias.

On this analysis of the context of dialogue for political debating, it cannot be taken for granted that the *argumentum ad populum* is always or inherently fallacious in political debate, based on the presumption of an illicit shift from critical discussion to negotiation, for negotiation is inherently a legitimate part of the context of dialogue. Therefore, to a certain extent, and in its proper place, interest-based argumentation of the *ad populum* type can be reasonable, tolerable, or at least not regarded as fallacious *per se*.

Hence, the dialectical shift theory does not justify calling the *ad populum* a fallacy whenever or wherever this type of argumentation occurs in political debates or speechmaking. On the basis of this theory, *ad populum* arguments cannot be condemned as inherently fallacious.

Still, the dialectical shift theory may not be useless. Dialectical shifts are important in understanding how *ad populum* argumentation works in everyday arguments like political debates. *Ad populum* can be a legitimate kind of argumentation, according to the shift theory, but it can still go seriously wrong or be unfairly used as a tactic of deception in some cases. And perhaps the shift theory can help in understanding what is wrong with it when it does go wrong.

More positively, the dialectical shift theory can be directly applied in the case where the context of dialogue is supposed to be that of a critical discussion, but where there has been an unlicensed or illicit shift to interest-based negotiation dialogue during an argumentation sequence. Of course, political argumentation is perhaps not very often like this, or properly so interpreted. But even so, the dialectical shift theory may help explain why *ad populum* argumentation is fallacious in some cases.

The general problem remains, however, of determining in a given case whether the use of an *ad populum* argument is fallacious or not. The shift theory does not solve this general problem for all cases.

A Classic Case

A classic case of the *argumentum ad populum* in political debate occurred during the leadership debate televised by the Canadian Broadcasting Corporation on 25 October 1988, just before the federal election. In that election the campaigning turned out to be a one-issue affair, centering around the proposed Canada-U.S. Free Trade Agreement. During the portion of the debate transcribed below, a journalist, Pamela Wallin, put questions to Prime Minister Brian Mulroney and to John Turner, leader of the Liberal opposition party. During this particular segment of the debate, the question put by Ms. Wallin was the following: "What kind of specific rules could you envision to ensure that *x* number of jobs stay in this country under Free Trade if that arrangement goes ahead?" Mr. Mulroney and Mr. Turner started out by discoursing on the pros and cons of how the Free Trade Agreement might affect various sectors of the economy and how each of the parties would deal with these effects.

Then at one point in the debate the two participants started to attack each other personally, in an emotional and more aggressive manner. This prelude built up to a spectacular interlude where both parties burst into a memorable emotional *ad populum* appeal, one after the other.

Case from the 1988 Canadian Federal Election Debate (FED)

[1] Mr. Turner: I think that the issues happen to be important for the future of Canada. I happen to believe that you've sold us out. I happen to believe that once you've entered . . .

[2] Mr. Mulroney: Just a second. You do not have a monopoly on patriotism, and I resent the fact of your implication that only you are a Canadian. I want to tell you that I come from a Canadian family and I love Canada, and that's why I did it—to promote prosperity— and don't you impugn my motives or anyone else's!

[3] Mr. Turner: Once a country yields its energy . . .

[4] Mr. Mulroney: We have not done that.

[5] Mr. Turner: Once a country yields its agriculture . . .

[6] Mr. Mulroney: Wrong again.

[7] Mr. Turner: Once a country opens itself up to a subsidy war with the United States on terms of definition . . .

[8] Mr. Mulroney: Wrong again.

[9] Mr. Turner: . . . then the political ability of this country to sustain the influence of the United States—to remain as an independent nation—that is lost forever. And that is the issue of this election.

[10] Mr. Mulroney: Mr. Turner—let me tell you something sir—this country is only about one hundred and twenty years old, but my own father, fifty-five years ago, went himself, as a laborer, with hundreds of other Canadians, and with their own hands, in Northwest Quebec,

built a little town, and schools, and churches. And they, in their own way, were nation-building—in the same way as the waves of immigrants from the Ukraine and Eastern Europe rolled back the prairies and, in their own way, and in their own time, were nation-building because they loved Canada. I today, sir, as a Canadian, believe genuinely in what I am doing. I believe it is right for Canada. I believe that, in my own modest way, I am nation-building, because I believe this benefits Canada, and I love Canada.

[11] Mr. Turner: I admire your father for what he did. My grandfather moved into British Columbia. My mother was a miner's daughter there. We're just as Canadian as you are, Mr. Mulroney, but I'll tell you this: you mentioned the hundred and twenty years of history. We built a country, east and west and north. We built it on an infrastructure that deliberately resisted the continental pressure of the United States. For a hundred and twenty years we've done it. With one signature of a pen, you've reversed that, thrown us into the north-south influence of the United States, and will reduce us, I'm sure, to a colony of the United States, because when the economic levers go, the political independence is sure to follow.

[12] Mr. Mulroney: Mr. Turner! With a document that is cancellable on six months' notice? Be serious!

[13] Mr. Turner: Cancellable? You're talking about our relationship with the United States. Once that document . . .

[14] Mr. Mulroney: A commercial document that's cancellable on six months' notice.

[15] Mr. Turner: A commercial document?

[16] Mr. Mulroney: It relates to treaties.

[17] Mr. Turner: It relates to every facet of our life. It's far more important to us than it is to the United States. It's far more important . . .

[18] Mr. Mulroney: Mr. Turner, please be serious.

[19] Mr. Turner: Well, I am serious. And I've never been more serious in my life.

The outbreak of *ad populum* posturing started by Mr. Mulroney in speech 2 is incited by Mr. Turner's personal attack, alleging that Mr. Mulroney has "sold us out" ("us" being Canadians) to the United States. Mr. Turner is claiming by implication that Mr. Mulroney is a kind of traitor to his country, by virtue of his having "sold" Canada to the United States, by signing an agreement that will destroy the ability of Canada to "sustain the influence of the United States" or to "remain as an independent nation" (speech 9). This barrage is an aggressive personal attack that shifts a heavy burden of presumption onto Mr. Mulroney to respond strongly or appear guilty and evasive.

Hypothetically, Mr. Mulroney had a choice of responses to this attack. He could have attacked the argument that the Free Trade Agreement is a "sellout" to the U.S., by posing critical questions to Mr. Turner's claims concerning the various consequences of the agreement. Or he could have responded in a more personal, emotional way, by affirming his own sincerity and the purity of his motives as a patriotic Canadian. Realistically, however, Mr. Turner's attack leaves no option but the second. By using the phrase "sold us out," Mr. Turner affirms his own solidarity with the Canadian people while labeling Mr. Mulroney an outsider, someone who stands outside the group solidarity and interests of the Canadian people, someone opposed to the group.

Looking carefully at the text of the debate, it is hardly surprising that Mr. Mulroney chose to reply with a personal *ad populum* appeal in the form of an affirmation of his patriotic sentiments. True, his patriotic appeal seems heavy-handed. He seems to be laying it on a little too thickly, saying that he loves Canada, that he comes from a Canadian family (speech 2), and, later, that his father was an immigrant laborer who helped to settle a small town in Quebec. But in the context of Mr. Turner's attack on his loyalty as a Canadian citizen, this emotional "touch a leaf" violin-playing session is not too surprising.

What is even more amusing, however, is Mr. Turner's follow-up by playing the same tune. In speech 11 he goes so far as to say that his mother was a miner's daughter. Tacitly he recognizes that Mr. Mulroney's *ad populum* overture was so successful that he must have a part in it too or at least respond with a few good licks himself.

The obvious criticism of both Mr. Mulroney and Mr. Turner's responses that the current textbook treatments would encourage students to consider is the following. Mr. Mulroney has committed the *ad populum* fallacy because instead of presenting evidence that would support his contention that the Free Trade Agreement will have good consequences for Canada, he launches into a gut, emotional *ad populum* appeal, presenting himself as the son of a humble Canadian laborer who helped build Canada as a nation. According to the same evaluation, Mr. Turner too has committed the *ad populum* fallacy. Instead of pointing out correctly that Mr. Mulroney has evaded the relevant issue of the consequences of free trade, Mr. Turner launches into an equally emotional, personal *ad populum* appeal, arguing that his grandfather was a settler in British Columbia (speech 11) and so forth. Both are trying emotionally to affirm their solidarity with the Canadian public instead of debating the issue of the projected positive versus negative consequences of the proposed Free Trade Agreement.

There is something in this criticism, for in fact none of the parties in this election did an adequate job of informing the public about the particulars of the Free Trade Agreement. But, on the other hand, there is something naive about this way of evaluating the FED case as a simple or obvious instance of the *ad populum* fallacy.

The specific question put by Ms. Wallin for this section of the debate concerned the number of jobs that would stay in Canada as a result of the Free Trade Agreement. These emotional, personal *ad populum* interludes by both participants did not address that question and were irrelevant to answering it.

On the other hand, it should be remembered that the context of this case is that of a leadership debate. Each of the participants has the quite legitimate goal in the debate of showing that he is the person that should be elected as leader of the governing party for the coming term. Questions of personal sincerity and who best represents the basic values of the Canadian public should be regarded as relevant to this debate. It is part of the democratic system to elect candidates on trust and to take their personal conduct, values, and motives into account in judging their suitability for office. Ethics and questions of personal character are therefore, to a reasonable degree, relevant.

One, therefore, has to look at the context of dialogue as a whole in reaching a balanced and fair evaluation of whether an *ad populum* fallacy has been committed. Was the personal, emotional appeal to popular sentiment excessive? Did it go beyond the bounds of relevance,[1] or was it

---

1. As Scott Jacobs pointed out in discussion at NIAS, Mr. Turner's initial attack was not purely on the consequences of Mr. Mulroney's policies on the Free Trade Agreement for Canada. It also concerned Mr. Mulroney's own personal *ethos*, his patriotic motives in supposedly "selling out" to the U.S. interests. Therefore, it was relevant for Mr. Mulroney to respond by citing his

an unfair tactic that went against the legitimate goals of the dialogue? These are the questions that need to be addressed, judging from the text of discourse given in a particular case.

In evaluating the FED case, it has become clear that there are reasons why a rational critic should not acquiesce too easily in the superficial claim that Mr. Mulroney committed an *ad populum* fallacy simply because he appealed to popular sentiments of Canadian patriotism in the debate.

Political Discourse

There are different kinds of *ad populum* arguments in political discourse, and these are arguments that involve different standards to judge them as correct or incorrect. One is the simple *ad populum* argument "Most of the people (voters, constituents) believe that policy *x* is right, therefore I support and advocate policy *x*." This type of appeal to popular sentiment is not inherently fallacious, but it is a simplified inference that can go wrong easily. The following discourse is part of a debate on the topic of capital punishment. (*Canada: House of Commons Debates*, 1981, June 11, 10521).

Case 3.2

> Mr. Bud Bradley (Haldimand-Norfolk): I rise to speak on this motion, Mr. Speaker, in respect of my own views and in respect of the views of my constituents. I personally favour some form of capital punishment. I feel in my heart that it is a deterrent. However, I was also elected by the constituents of Haldimand-Norfolk to represent them and their views in this House.
>
> In a recent questionnaire to my constituents, one of eleven varied questions read:
>
> Do you feel that capital punishment should be reinstated: Yes; No; Undecided.
>
> In response to that questionnaire, 85.6 per cent indicated yes, 5 per cent were undecided and only 9.4 per cent indicated no. I doubt if any constituency varies greatly from others in this country.

---

patriotic motives instead of sticking purely to the issue of the Free Trade Agreement and its consequences.

People in Haldimand-Norfolk have the same concerns as others across this land: concern for their children, concern for their friends and relations, and that concern is growing.

There has been much talk this evening about statistics on murder. In fact, the number of attempted murders has increased tremendously. There are other factors affecting the number of murders committed. Consider such things as improved equipment and improved skills of medical personnel such as those the hon. member for Hamilton West (Mr. Hudecki) possesses. I am sure he is well aware that countless lives of attempted murder victims can now be saved which possibly could not have been saved 20 years ago. This is evidence we must consider.

We live in a democratic society where the majority should rule. I suggest that the people's voice should be heard. Abolition has failed to convince them. People are demanding a return to capital punishment. We must be a responsible government. We must be mindful of the wishes of our people or we will have failed.

This argument presents a straightforward appeal to popular opinion. Mr. Bradley states that he himself is in favor of capital punishment. But his main argument, in the discourse above, for adopting a *pro* capital punishment position is that his constituents have been shown to be overwhelmingly in favor of it. To support his argument, Mr. Bradley actually goes so far as to quote the results of a poll of his constituents.

Is Mr. Bradley's *ad populum* argument fallacious? It would seem not. In a democratic system, it is quite appropriate for an elected officeholder to be "mindful of the wishes of the people." Moreover, Mr. Bradley is not dogmatic or excessive in appeal to popular sentiment. He claims that the voice of the people "should be heard," not that it should require a return to capital punishment.

What is especially interesting to note in this case is that appeal to the people should be a defeasible kind of argumentation that should not be the only consideration in arriving at a policy for action. It is proper for a politician to go against the majority if he or she feels strongly that the majority is wrong on a particular issue. Indeed, politicians are expected to go against the received opinions in some cases, even at some cost in votes or other forms of support. A stand against popular pressures can be heroic and courageous. Thus, a reflexive or dogmatic submission to popular opinion could, in some instances, be rightly regarded as an *ad populum* argument that is open to serious criticism.

Note, however, that in this case, there is no evidence that Mr. Bradley has made a misjudgment or argued in a fallacious way. He uses the *ad populum* argument explicitly, but that in itself does not appear to be fallacious. The use of the argument from popularity is straightforward, but not unreasonable or fallacious.

In many other cases in political argumentation, the use of the *ad populum* is less straightforward and much more subtle. In these cases, the evoking of popular sentiment is a matter of "style" and "image," of how a candidate or officeholder presents himself to the public. Public relations experts expend considerable time and money devising strategies to build an image for their client as a "man or woman of the people." In cases where the candidate comes from a privileged, upper-class background, their job is not too easy.

George Bush was a graduate of Andover and Yale, and also the son of an elite and wealthy family—an outstanding, "preppy," Establishment figure. His campaign manager, Lee Atwater, had to do some "serious retrofitting" for the Bush election campaign, based on "an emotional, populist appeal to traditional values (of flag, family, and security)" (Fineman 1988, 58). The strategy was to use those themes to secure a base in the south, according to Atwater. Bush speeches were tailored to fit this *ad populum* strategy.

Case 3.3

> At his rallies, small-town folk boo and laugh when he mentions the hated "liberals." He warns that "they" will invade people's privacy and raise taxes. His foes, he says, want to concentrate power in Washington, an incubator of misguided "socialist" thinking. He blasts the "elitists" at Harvard and, as part of a strident law-and-order appeal, poses with uniformed cops at the drop of a brass shield. He says the flag must be honored again, the death penalty restored. He likes to visit state fairs, bass-fishing contests and Roman Catholic churches in ethnic neighborhoods. (Fineman 1988, 58)

What is to be said about this type of *ad populum* appeal in political speechmaking? In this case, there does seem to be something contradictory or questionable about it; a patrician, Establishment figure from a wealthy family is being portrayed as a folksy populist who likes pork rinds and the Oak Ridge Boys. Isn't this really a clever deception?

Perhaps it is, to some extent, but it appears that all presidential candidates do something like this, more or less successfully, in order to secure a broad enough base of political support for their program. True, it is a manipulative tactic. But, on the other hand, it does seem to be a necessary part of a political system where voters are going to make choices based on how a candidate is personally portrayed in the media. Thus, the campaign managers must compete in open argumentation, a kind of free marketplace where the arguers do their best, and it is up to the voters to decide who made the best case for their candidate and the candidate's policies. To condemn such an argument as fallacious, therefore, seems too strong and even inappropriate as a criticism.

It is one thing to be critical, or even suspicious, about this kind of rhetoric in political speeches and messages. It is something else again to show, in a given case, that the use of such an *ad populum* appeal is fallacious. Much depends on how the appeal to popular sentiment is used in the given case.

A nice distinction advocated by Brinton (1988, 211) can be applied to contrast the uses of popular opinion in cases 3.2 and 3.3. Brinton calls the attempt to arouse emotion, say, in a crowd, an *evoking* of emotion. Whereas an appeal to emotion as a basis for action could be described as an *invoking* of emotion. The attempt to arouse emotion, or appeal toward emotion, is an evoking of an existing feeling or emotional predisposition in an audience. This evoking of emotion was the tactic used by the Bush campaign strategists, who were trying to create an aura of popular approval around their candidate by appealing to commonly accepted tokens of values and sentiments as props to convey a feeling of folksiness.

On the other hand, sometimes emotion is appealed to as a basis for action in practical reasoning or in deliberation on how to proceed with a debatable course of action. Practical reasoning is based on premises that describe an agent's goals, and on other premises that describe ways the agent knows to carry out these goals in a given situation (as the agent sees it). Practical reasoning is characteristically based on presumptions concerning a particular set of given circumstances of an agent (Walton, 1990).

In political deliberations that concern a controversial course of action or proposed legislation, there are generally two opposed parties engaged in *pro* and *contra* argumentation on the best course of action or policy to follow. In such a case, according to the model of political debate proposed by

Windes and Hastings (1965, chapter 7), the two opposed advocates argue for the "positive" consequences of their own proposals and against the "negative" consequences of the proposal put forward by the other side. But they also argue about the goals and values of the public or nation on whose behalf the proposal is intended. Thus, according to Windes and Hastings (1965, 213), the value system of the culture or national group, the "beliefs of the audience," must be taken into account by any advocate in political debate building a case for a proposal relating to group or national political action. In such a case, what Brinton (1988, 211) calls the invoking of emotion—in this case, popular sentiment—can be a legitimate part of the practical reasoning that is a basis for action.

Epideictic Speeches

In some of the best-known cases of political argumentation, the aim is not to resolve a conflict of opinions on some disputed issue of current public policy or need for action. Instead, a single speaker, on some ceremonial occasion, like a funeral, a memorial day, or a special holiday, delivers a stirring emotional speech whose aim is to express, solidify, or reaffirm group spiritual values in an expressive and aesthetically pleasing way. This so-called *epideictic* type of speech was recognized by Aristole in the *Rhetoric* as a distinctive and important genre of persuasive discourse. The Greek sophists were expert at performing this kind of speech.

Perelman and Olbrechts-Tyteca (1969, 48) have described the aims and methods of epideictic oratory. A single orator is to make a speech, not on a topic that is of urgent public concern, but on a more abstract, general theme, usually one that has some special emotional impact as praise of some value that is dear to the audience. Such a speech is regarded as a showpiece, and its intent is more dramatic than practical. The goal is not to prove something, as in an inquiry or investigation into a subject, but rather to enhance some existing value professed by the audience to whom the speech is directed. The audience is not meant to reply or to debate with the speaker. After the speech is finished, the audience can leave, without responding to it in any visible way. The speech can also be written and circulated in printed form or replayed in film or video form.

Perelman and Olbrechts-Tyteca (1969, 51) described the essential aims and elements of the epideictic speech by specifying the speaker's goals.

Unlike the demonstration of a geometrical theorem, which establishes once and for all a logical connection between speculative
truths, the argumentation in epideictic discourse sets out to
increase the intensity of adherence to certain values, which might
not be contested when considered on their own but may nevertheless not prevail against other values that might come into conflict
with them. The speaker tries to establish a sense of communion
centered around particular values recognized by the audience, and
to this end he uses the whole range of means available to the
rhetorician for purposes of amplification and enhancement.

In order to establish this "sense of communion" between speaker and
audience and to increase the "intensity of adherence" to group values, the
speaker must presumably make emotional appeals to stir the audience.
Above all, the speaker must appeal to the popular sentiments of the
audience and focus those sentiments into a stirring reaffirmation of the
values of the audience as a group.

The goal of the epideictic speech as a type of discourse is—using
emotional means, including poetic and appealing language—to awaken or
reaffirm a feeling of group solidarity or communion. Clearly then the
argumentum ad populum, far from being a fallacy in this type of dialogue, is
really the central technique of argumentation around which the whole
speech is built. The *ad populum* appeal does not go against the goal of the
dialogue. It is the means used to achieve the goal.

Any *ad populum* appeal in argumentative discourse is therefore not
automatically fallacious. An *ad populum* appeal in a political speech, for
example, may look like a fallacy, but if that political speech is supposed to
be an epideictic discourse, the presumption that any *ad populum* (or other
type of emotional appeal in it) is irrelevant or beside the point may be
questionable to say the least.

Consider the speech "For the Freedom of Man We Must All Work
Together" delivered by John F. Kennedy at the inauguration ceremony on
20 January 1961 in Washington, D.C. In this famous speech, reprinted in
Vital Speeches, Mr. Kennedy appealed to many emotions. He appealed to
pity for "those peoples in the huts and villages of half the globe struggling
to break the bonds of mass misery" (226). He appealed to solidarity:
"United, there is little we cannot do in a host of new co-operative
ventures" (226). And he appealed to the threat of force, saying, "Let every
other power know that this hemisphere intends to remain master of its own

house" (226) and "In the past, those who foolishly sought power by riding the back of the tiger ended up inside" (226). Should these emotional appeals be labeled with the classifications of the fallacious *argumentum ad misericordiam*, *argumentum ad populum*, and *argumentum ad baculum*? They certainly appear to fit the standard descriptions of these traditional fallacies.

Before agreeing to any of these classifications, however, one should ask what the context of dialogue for Mr. Kennedy's speech was supposed to be. Was it a critical discussion of national or foreign policies? Was it an inquiry into some matter in order to prove something, some particular conclusion, to the audience? If either of these characterizations of the speech were accurate, a good case could be made for the inappropriateness or irrelevance of the emotional appeals made in the argumentation.

Both of these interpretations are quite implausible, however. It is clear that this discourse was meant to be an epideictic speech. The speech does sketch out a very general basis for foreign policy, but it is clear that it was not meant to be a statement of specific domestic or foreign policies. It is meant to be an affirmation of Mr. Kennedy's own values as an American and also an appeal to all Americans to join with him in an affirmation of loyalty to these values. In the most memorable part of the speech, Mr. Kennedy said, "And so, my fellow Americans: ask not what your country can do for you—ask what you can do for your country." The *ad populum* argument is the fabric that holds the whole speech together as a meaningful discourse—it is not an inappropriate or irrelevant appeal that hinders or runs counter to the goals of the dialogue. Here the *ad populum* argument is used to evoke popular sentiments in a way that supports the goals of the dialogue.

Many textbooks cite Mark Antony's funeral oration over the body of the murdered Caesar as the classic case of the fallacious *argumentum ad populum*. Castell (1935, 23–30) quoted this speech at length, concluding that the irrelevant appeals of popular sentiment in it are used by Antony to work the crowd into an emotional state where they exit, bent on the destruction of Brutus, one of the leading participants in the killing. According to Castell, the crowd knows "as little as before Antony spoke, of what it is all about," and the emotional appeal, while irrelevant to the real issue, has been effective in persuading the crowd. Hence the fallaciousness of the *ad populum*. Copi (1986, 97) cites this case as well.

But what is supposed to be the real issue of Antony's speech? This is something of a literary question, since the speech is part of a Shakespearean play. But presumably, the purpose is to give a funeral oration. Antony turns this to his own purposes, making it into a vehicle for something else. But for what? An epideictic speech to incite a mob? A kangaroo court to convict the murderers in their absence? Clearly there is a dialectical shift here, but the type of dialogue to which the argumentation shifts is perhaps not too clear. On the other hand, it is not by any means clear that a trial of some sort is supposed to be taking place. And therefore, it may not be altogether appropriate to criticize Antony's speech for evading the facts which would be relevant to proving Brutus guilty in a court of law.

Presumptive Inferences

One may be inclined to think that the inference "Everyone else is doing x, therefore I am going to do x" is the paradigm of the *ad populum* fallacy. But arguments having this form can, in some cases, seem quite reasonable. The reason is that presumptions concerning commonly accepted ways of doing something can be part of a legitimate basis for concluding to an action in some cases.

Case 3.4

> I am just getting off the train in the central station of a foreign city. I have never been to this city and do not know which direction the exit to the street is. Carrying my heavy bags, I step off the train and look in one direction, then the other. In both directions there are stairwells at the end of the long platform. However, nobody is going in one direction, and everyone pouring off the train is hurrying in the other direction. I follow the crowds in the second direction.

In this case I have to decide to go one way or the other. I could just choose at random, for I have no real "hard evidence" to go on. But I do have heavy bags, and if I pick the wrong direction, it could mean carrying these bags for a longer distance than is necessary. As a working presumption, in the absence of any better indications, it makes sense to follow the crowds.

What makes the inference to follow the crowds reasonable in this case is that the assumption that the crowds of people are heading in the right direction towards the exit is a defeasible presumption. I can presume that most of these people are familiar with this station and know the right direction to go in when they get off the train.

This presumption is not an appeal to expert opinion—linked with *ad verecundiam* argumentation—at least not directly, because it is misleading to say that these people are "experts" on this particular station. Instead, they are in a special position to know about the way to leave the station, or so I presume, because they have done it before or perhaps are with someone who has done it before, etc. But the presumption I act on is linked very closely with the *argumentum ad ignorantiam*, for in the absence of knowledge on whether to go this way or that, I have no way of knowing that the crowds are wrong, so I can act on the tentative presumption that they are right. It is a matter of shifting the weight of presumption in one direction instead of another, based on a need to act one way or the other. The slight weight of presumption on one side of the decision tilts the balance of argument in that direction.

Thus the *ad populum* argument in this case is best conceived as a reasonable argument precisely because it is a weak argument that tilts a weight of presumption one way, in the absence of more definitive evidence on which to make a choice that has to be made, for practical purposes of action, one way or the other. This argument, as such, is not fallacious, although it could become fallacious if the circumstances of the case were altered.

Suppose, for example, that while following the crowd and carrying my bags, I encounter a sign reading, "Exit this way: sports stadium this way," where one arrow (in the sports stadium direction) is toward the direction the crowds are going and the other arrow points toward the opposite direction. Assuming I can interpret this sign clearly and have no reason to doubt the veracity of what is indicated, it now becomes reasonable to presume that these crowds are hurrying toward the sports stadium. Since I want to go toward the exit to the street and do not want to go to the sports stadium, by practical reasoning, I should draw the conclusion to reverse my direction, heading in the direction opposite to the way the crowd is going. This reversal presumes that the sign is better evidence than the direction of the crowds, given the new information about the direction of the sports stadium and the station exit.

One interpretation and reconstruction of the sequence of reasoning in this case is the following. First, I act on a weak and defeasible presumption. In the absence of knowledge and decisive evidence, and given the practical need to proceed one way or the other, this seems the most reasonable course of action. Therefore, even though the argumentation is *ad populum*, an instance of the argument from popularity, it is not fallacious. It is reasonable as a way of drawing a conclusion to action on the basis of a defeasible and weak, but relevant, presumption. Admittedly, it is a lack-of-knowledge (*ad ignorantiam*) inference, but presumptive reasoning is generally defeasible and tentative, a provisional basis for action that is sufficient to alter a burden of proof in the absence of better evidence or definite knowledge in a given case.

This acceptable type of *ad populum* reasoning can become fallacious, however, if I, the proponent, continue to cling to it, or to act on it dogmatically, even in the face of contra-evidence which I come to know about in time to influence my action. Suppose, for example, that even after reading and correctly interpreting the sign, I were to argue to my wife, "Well, look, I know the sign says the exit is that way. But all these people are going this way. All these people can't be wrong. They must be going toward the exit. Let's follow the crowd." This apparent dogmatic refusal to take the new evidence properly into account could be grounds now for alleging that I am committing an *ad populum* fallacy. My wife might reply, "Well, we can't ask anyone very effectively, because we don't speak the language here, and not many people speak English. But we do know that this sign means 'exit' this way. Why follow the crowd? It looks like they are going to the sports stadium." This seems like a reasonable line of argumentation, but I reply, "Everybody is going that way. They must be right. We must follow them." In so replying, I have shown that I am in the grip of some dogmatic preconception in favor of following what everyone is doing, even in the presence of good evidence (that I have not successfully rebutted) to the opposite conclusion. Here evidence has come in, but I am ignoring it and cleaving to my original presumptive conclusion dogmatically.

In the proposed extension of case 3.4 above, then, there is evidence from the text of discourse and the context of dialogue that the *argumentum ad populum* is being used in a way that could rightly be criticized as fallacious. But note that it is not the *ad populum* argument *per se* (as used in case 3.4) that is fallacious. It is the abuse of the *ad populum* argument that results from leaning on it too hard, even in face of contrary evidence. It is a presumptive inference that cannot bear this weight, once the dialogue is extended and new evidence is introduced that makes the case more complicated.

Presumptive inferences, then, are defeasible; they may have to be given up if new evidence comes into a dialogue that rebuts them. To take the dogmatic line of refusing to treat such an inference as defeasible can be considered erroneous or even fallacious reasoning. To push ahead too strongly with an *ad populum* argument, to use it to defeat the introduction of any new argumentation or evidence that comes into a discussion, is fallacious argumentation.

To conclude: the *argumentum ad populum* is an inherently reasonable kind of argumentation that only goes wrong, or is fallacious in some cases, where it has been used improperly. It should not be presumed that this kind of argument is generally fallacious. Instead, the burden of proof should be on the critic to show why it has been used fallaciously in a particular case.

In the everyday practices of critical discussions on controversial issues, initial presumptions are often based on popular opinion. This weight of popular opinion provides an initial horizon that may be significant in setting burden of proof and in providing an initial weight of presumption an argument may have to overcome or deal with. An argument should be judged strong or successful in a critical discussion, at least partially, in relation to how it alters this initial weight of presumption. In some cases a really enlightening, successful, and surprising dialectical argument goes against popular opinion yet is hard to refute because of the logical argumentation behind it.

Thus, it is easy to fall into a confusion regarding the worth of popular opinion in argumentation. As an initial basis for the horizon of presumption in the opening stages of a critical discussion, popular opinion should properly carry some weight. But popular opinion should not be the last word, the only criterion, in judging which side of a disputed issue has made the stronger case in a critical discussion. It is just one factor among others to be evaluated. But it is easy to think of popular opinion in all-or-nothing terms, to think of it either as a worthless and unreliable (even fallacious) presumption in argument, or alternatively as a final arbiter or criterion that is not open to challenge or rebuttal. This black-and-white approach tends to make us suspicious about appeals to popular opinion in argumentation.

When Is It a Fallacy?

The five major reasons given by the logic textbooks for condemning *ad populum* argumentation all have something in them. The *ad populum* is a tactic that is open to criticism as a bad move in argument because it does, in some cases, appeal to emotion or popular prejudice in order to introduce irrelevant factors to evade meeting a proper burden of proof in a critical discussion. And the argument from popularity is sometimes leaned on more heavily than it can bear. But none of these errors or dubious moves in argumentation are necessarily fallacious. A fallacy is a serious, systematic, underlying error of argument that justifies strong refutation of the argument. To claim that an argument commits a fallacy is a serious charge that carries with it a burden of proof on the critic.

In many cases there is a danger of critical failure associated with *ad populum* arguments, but the appeal to popular sentiment is not so bad that it should be called fallacious. In the FED case, for example, the appeal to patriotic sentiments was something of a waste of time because it would have contributed more to the debate to talk about the Free Trade Agreement specifics. But it was not altogether clear that Mr. Mulroney committed a fallacy.

In a common case like 3.3, where a political speaker evokes an aura of social alignment with popular values by folksy rhetoric and appeals, the problem is not so much a fallacy as simply an unsuccessful or phony attempt to project a popular image. It is acceptable in political campaign speeches in a democracy for a speaker to project a positive *ethos* by identifying himself or herself with popular values. But when a patrician politician like George Bush tries to portray himself as "one of the boys" in a heavy-handed or obviously insincere or inept manner, this is more a failure to project a popular *ethos* than a fallacy.

Other faults of argumentation may perhaps be more successfully identified with *ad populum* fallacies.

Ad populum arguments can be criticized for specific faults and failures in cases where the *ad populum* argument has been used to block or hinder legitimate goals of a dialogue.

In one case the appeal to popular opinion ought simply to be defeated by other factors or presumptions that outweigh it.

Case 3.5

A Roman citizen and his neighbor are having an argument about whether it is right to own slaves. The time of their conversation is 100 B.C. The neighbor argues: "Everybody owns slaves, therefore it is all right."

Here the argument appears weak and unconvincing because the appeal is to a common practice based on discriminatory subjugation. Owning slaves was then common, but there are reasons for concluding that it is not morally acceptable. By current popular opinion, this argument would be strongly rejected. From the vantage point of 100 B.C. Rome, this argument was (no doubt) persuasive. But public opinion changes. An *ad populum* argument that was powerful in one climate of opinion becomes no longer persuasive from a different point of view.

The problem in this case, however, is not that an *ad populum* fallacy has been committed. It is simply a weak and unconvincing argument because the weak weight of presumption brought in by popular opinion ought to be rebutted by other factors. The appeal to popular opinion is unconvincing, by current standards of popular opinion. It was based on a point of view that, as can be seen now, was wrong, absolutely.

In another case, the premise containing the appeal to popular opinion is convincing enough, but the conclusion derived from it does not follow.

Case 3.6

> The Golden Rule is a fundamental principle of every system of ethics ever constructed. Everyone accepts it as an ethical principle. Therefore, the Golden Rule is undeniably correct as an ethical principle.

Here, the *ad populum* argument carries some weight. If every system of ethics acknowledges the Golden Rule, then it must be an important principle for ethics, and there is some reason to presume that it is an acceptable principle. But it does not follow that it is "undeniably correct" as an ethical principle. This strong conclusion pushes the weight of presumption a bit too far, excluding further discussion of the reasons supporting the Golden Rule as an ethical principle.

The problem in another case is the weight of presumption if an *ad populum* argument is pushed too hard, but in a different way.

Case 3.7

> The people of this city have always chosen the finest citizens of their community to represent them. This fact is born out by the long record of intelligent and fair decisions reached in this room. I am sure that you will want to add my proposal to that commendable list.

This argument is open to critical questioning because the wisdom of past decisions and the personal characteristics of the councillors are not much of a weighty factor in deciding on the merits or faults of the specific proposal at issue. The argument omits any serious evidence and instead tries to flatter the councillors. The argument given is, in effect, "You men and women are fair and intelligent, therefore, accept my proposal." This argument is very weak, at best. A more appropriate and relevant argument would be "My proposal is fair and intelligent, for the following reasons; therefore accept my proposal." The given argument is deficient because of what it does not say.

The other problem with the argument in this case is that the proponent is trying to browbeat the respondents by saying, in effect, "If you do not vote for my proposal, you will have shown that you are unfair and unintelligent, and you will go against the pattern of your previous voting record." This is an indirect form of *ad hominem* argumentation that puts an undue pressure on the respondents.

Another case is somewhat similar in that the proponent indirectly puts pressure on the persons to whom the argument is directed in a way that tries to block off discussion.

Case 3.8

> Every loyal American will stand behind this proposal and back it as fully as possible.

This *ad populum* argument tries to cut off dialogue by excluding any American who does not back the proposal as disloyal. Presumably, the audience to whom the argument is directed is composed of patriotic Americans. Hence, the use of this *ad populum* leaves them no leg to stand on. Anyone who attempts to criticize the proposal is immediately categorized as a "disloyal" American. This indirect *ad hominem* strategy effectively cuts off any possible attempts to criticize the policy.

Bailey (1983, 134) quoted Walter Reuther speaking in 1957 on the subject of racketeers in trade unions.

Case 3.9

> I think we can all agree that the overwhelming majority of the leadership of the American movement is composed of decent,

honest, dedicated people who have made a great contribution involving great personal sacrifice, helping to build a decent American labor movement. . . . We happen to believe that leadership in the American movement is a sacred trust. We happen to believe that this is no place for people who want to use the labor movement to make a fast buck.

The problem here, as pointed out by Bailey (134), is that the repeated use of the phrase "We happen to believe" frames what Reuther says as a value of the group that is beyond all doubt and questioning by any member of the group. The universe is divided into believers and nonbelievers, and the "we," in contrast with the nonbelievers, are placed on the high moral ground. Bailey (135) calls this strategy "the rhetoric of belonging." The nonbelievers are the people who "want to use the labor movement to make a fast buck." These people are excluded from the audience of true believers in the movement whose values and loyalty are appealed to and reaffirmed by the speech.

In constructing or laying out a line of argument to convince a respondent to act in a particular way, or to undertake commitment to a point of view, popular opinions, or accepted, routine ways of doing something, may be very good places to start in finding initial, working presumptions as premises. Typically, practical reasoning works from existing premises that are reasonable presumptions in a situation but that may need to be refined or altered as the agent's knowledge of the situation improves. There is nothing wrong with this way of building a case *per se*. There is nothing wrong with accepting or proposing premises based on popular opinions or sentiments.

Of course, it must be remembered that all such provisional presumptions are defeasible. If good evidence of a contrary nature comes forward, a participant in argument should have the right to raise critical questions challenging or even refuting his or her own or the co-arguer's initial presumptions. Such presumptions being defeasible, putting them forward too aggressively or clinging to them dogmatically can be faulty or even fallacious argumentation.

But appeals to popularly accepted practices or commitments has positive, constructive uses in argumentation. Emotional appeal, in such cases, may have more than just a monitoring or tutoring function. Popular opinion can be a legitimate baseline premise crucial in supporting a line of argument needed to build a case. It is necessary in dialectical argumentation generally, but especially in practical reasoning in political argumentation.

The *argumentum ad populum* is most effective as a technique of argumentation in establishing a link—an identification or social alignment—between the commitments of the speaker and the audience. The speaker may not be able to prove all the premises needed to make a case in favor of a proposal put before a particular audience, and such an exhaustive approach to fulfilling the burden of proof may be neither necessary nor effective. If, instead, the speaker can get the audience to go ahead on the presumption that their commitments on the issue are pretty much the same as those of the speaker, he or she can cover a lot of ground in argumentation quickly. The message is "Trust me; I'm one of you." If this message can be effectively put in place by a speaker, it can obviate the need for answering a lot of critical questions that otherwise might be raised by the audience before granting their approval of a proposal.

As a tactic of argumentation the *argumentum ad populum* exploits the link between a proponent and a respondent, presuming they have much in common in their basic positions. The proponent is arguing, in effect, "There is really no need for me to prove all this, because you and I are in basic agreement on it." By putting a blanket presumption in place, the proponent is asking to be relieved of a detailed burden to prove the individual points required to make a case. This way of arguing is not an inherently fallacious or wrong way to proceed, but naturally, it tends to be a sweeping presumption that is open to abuses and deceptive manipulations.

Whether a given use of the *ad populum* argument is fallacious or not rests on how the presumption is advanced in the sequence of dialogue exchanges between the participants in the type of dialogue that they are supposed to be engaged in. Putting a presumption forward without proving it is acceptable *per se*. But if the respondent raises legitimate critical questions or brings forward evidence that throws the presumption into doubt, the proponent must not push ahead too strongly by trying to block this questioning or by trying to prevent it from arising, by shutting down the proper avenues of further discussion. Where this overly aggressive tactic of appealing to a popular opinion or sentiment is used to block or hinder the legitimate goals of a dialogue, it is proper to allege that a fallacy of *ad populum* has been committed.

In cases 3.6 through 3.9 it is appropriate to bring forward the charge that the proponent has committed the *ad populum* fallacy. In case 3.5 the failing is a common error, to be sure, where a proponent tries to argue that some practice is morally acceptable because the majority do it or approve of it. But in this case the weight of presumption put forward on the basis of common practice or acceptance of a policy at one time is quite strongly and decisively outweighed by countervailing evidence or reasons that nearly everyone at a later time would readily accept as overruling. This is not a fallacy, even though it is such an unpleasant or distasteful argument that one is inclined to strongly condemn it as "fallacious." Not every bad argument that appeals to popular opinion or sentiment is a fallacious *argumentum ad populum*, according to the analysis presented above.

4

ARGUMENTUM AD MISERICORDIAM

The textbooks are divided on the question of whether it is worthwhile treating the appeal to pity (*argumentum ad misericordiam*) as a special category of fallacy. Copi (1986) and many others include it, but an equal number make no mention of it as a separate fallacy in its own right.[1]

Hamblin (1970, 43) mentions the *ad misericordiam* only briefly, doubting whether it is really a fallacious type of argument to the extent presumed by the standard treatment of the textbooks.[2] Hamblin notes that in a lawsuit or a political speech, propositions are often put forward in argument as

---

1. The author has not attempted a systematic survey—this estimate is only a guess, based on experience and checking a dozen or so textbooks.

2. It is not known where the *ad misericordiam* originated as a fallacy, but of course Aristotle did write extensively about *pathos* as a means of persuasion in the *Rhetoric*, including a discussion of pity. See Brinton 1988.

guides to action, and "where action is concerned, it is not so clear that pity and other emotions are irrelevant."

The *ad misericordiam*, then, poses a problem for the analyst of fallacies. Is it a distinct fallacy in its own right, as opposed to just another way that an argument, by playing on emotion as a distraction, can fail to be relevant? And if it can be a reasonable argument in some cases, how can one judge that it is fallacious when it is fallacious? This is the problem of "pinning down" the fallacy.

The approach to the *ad misericordiam* in this chapter will be first of all to understand why and how it can be used in some contexts as a reasonable type of argumentation. By coming to understand how and why it is such a powerful type of argument, one which can exploit pictorial appeals to sympathy with overwhelming impact, it will be possible to study the problem of how best to react to it. Finally, five gradations of reaction in evaluating *ad misericordiam* arguments will be distinguished, implying that the standard treatment is too simplistic to be useful.

Textbook Accounts

Some of the textbooks distinguish between fallacious and nonfallacious appeals to pity by making a positivistic dichotomy between facts and sentiments. If an argument is only about facts, then sentiments are irrelevant. But if an argument is about sentiments, then sentiments are relevant.

According to Copi (1986, 95), *ad misericordiam* is "the fallacy committed when pity is appealed to for the sake of getting a conclusion accepted, where the conclusion is concerned with a question of fact rather than a matter of sentiment." According to Freeman (1988, 74), the arousing of pity "does not guarantee that we have a fallacious appeal to pity," but such an appeal is fallacious "when factual considerations are relevant." In a given case, Freeman concludes, the appeal to pity is fallacious if it cannot be determined whether the emotion was appropriate or not.

But is it possible to determine in a given argument whether the argument is partly about sentiments or whether only "the facts" matter? Perhaps in some cases, this can be done. In a charitable appeal, sentiments are clearly relevant and important. In a scientific inquiry—for example, in geometry or physics—where a proof is being given or an experimental result presented, sentiments and appeals to pity are clearly of no evidential worth. But what about a critical discussion on some subject of ethical controversy, like abortion or euthanasia (see cases 4.6, 4.7, and 4.8 below)? Are facts all that count here, or do sentiments have some place in the arguments? In such a case, the criterion that calls for a positivistic bifurcation falls down. Sentiments can be relevant, but they can also be played on excessively or in an inappropriate way.

Note that Freeman's allocation of burden of proof is too mild: if it cannot be determined whether an appeal to pity is appropriate or not, the burden of proof should be on the critic who alleges that the appeal is fallacious to show convincingly that the appeal was genuinely inappropriate. The positivistic approach yields a false optimism that facts are one thing, sentiments another, and that there is generally no problem of telling when sentiments are irrelevant. On the contrary, in a critical discussion this judgment requires a careful examination of the evidence from the text of discourse and the context of dialogue. The critic who alleges that an appeal to pity is fallacious must marshall the relevant evidence to properly substantiate such a serious claim by showing that the appeal is irrelevant or inappropriate.

Cederblom and Paulsen (1982, 100) propose that an appeal to emotion in argument is illegitimate if it presents a motive in place of support for a conclusion. When deciding how to act, they argue, all motives are, in a sense, relevant. But when deciding what to believe, they continue, motives are not relevant. This last claim appears dubious, however. If one is deciding whether or not to believe I have a certain motive, then my motives are relevant.

Many arguments are partly about beliefs and partly about actions, combining both concerns; neither belief nor motive is the central focus or goal of the argumentation. In a critical discussion on an ethical controversy like abortion or euthanasia, for example, the dialogue is concerned with the reasoned commitments of the participants. To a certain extent this concern is with acceptance of propositions as true or false, justified or unjustified, etc., but it is also with public policies and how people should act. Hence motives and sentiments may properly be involved in such arguments, although they may be appealed to in a fallacious or inappropriate way in other arguments.

Damer (1980, 88) rejects the simple bifurcation that appeal to pity is fallacious when it involves coming to a *belief* on the basis of that pity as opposed to evidence and nonfallacious when it involves coming to a decision about a course of action. Damer, rightly in this author's view, contends that appeal to pity in persuading someone toward a particular course of action can be nonfallacious.

Weddle (1978, 35) proposes that an appeal to pity is a relevant appeal if "presented with a force proportional to the issue's claim to consideration." Thus a fallacious appeal is one that "milks the issue" by going to excessive lengths, or using the appeal to pity in a heavy-handed way that tries to push it far beyond its reasonable weight as a claim to consideration.

Damer and Weddle take a less simplistic and positivistic approach to the *ad misericordiam* as a fallacy; their approach does not depend on an appeal to a sharp dichotomy between facts and values as the sole distinguishing criterion for judging cases. They both see appeal to pity as legitimate in many contexts of argument and see the fallacious appeal to pity arising out of specific faults where the appeal has been used inappropriately or laid on excessively. But by making the problem more subtle and sophisticated, they also make it more difficult to solve. How does one tell the difference between the fallacious and nonfallacious appeals to pity in argumentation?

The old, positivistic idea that some arguments are about facts and other arguments are about sentiments is not only based on an epistemological dualism that many consider untenable, but it is also of little or no help in important cases. Perhaps sentiments can be clearly and decisively excluded where the context of argument is that of a scientific inquiry. But even if pity is relevant, for example, in a charitable request or argument for action to help someone, the appeal to pity may be put forward in an inappropriate, heavy-handed, or fallacious way.

Much does depend, however, on the context of dialogue of the given argument. As stated above, an appeal to pity as an argument worthy of consideration and weight must be judged quite differently in a charitable appeal for help than in a scientific inquiry. In a critical discussion, sentiments as well as facts are important in fulfilling the goals of the dialogue. Facts make for a well-informed discussion and can be introduced through the device of an intersubjective testing procedure, or ITP (van Eemeren and Grootendorst 1984, 167). But sentiments are also important, because in order to find premises that represent real commitments the respondent will stick to and not retract easily, the arguer must judge by empathy what the respondent's basic position is. Conjectures must be made based on presumptive reasoning concerning the respondent's darkside commitments on the issue of the dialogue. This means articulating the feelings or sentiments of the respondent, based on how the arguer thinks he or she responds, as a person, to the issue.

What is a convincing argument for this respondent is therefore a matter, to some extent, of how the respondent feels about the issue. So sentiment is not irrelevant, in a critical discussion, to evaluating whether an argument is successful or not in relation to the goals of the discussion.

In many of the cases of *argumentum ad misericordiam* cited by the textbooks, the context of dialogue would appear to be that involving some sort of critical discussion of an issue. In such cases, it is by no means evident that appeal to pity can be banned as totally irrelevant. How does one determine fallaciousness versus non-fallaciousness then?

Reasonable Appeals to Pity

Can the attempt to arouse emotion in a discussion be a legitimate basis for advocating a conclusion to a line of argument? Brinton (1988, 211f.) has presented a case that would indicate that the answer should be "yes." In this case an agent of persuasion attempts to get a recipient of her argument to feel certain emotions in order to get her to do something.

Case 4.1

> You are my daughter. An aunt who has always taken a special interest in your well-being—has treated you with kindness, remembered your special occasions, helped with your educational expenses, and so on—has fallen on hard times and is in ill health and lonely and depressed. I remind you of her many kindnesses and present you with a detailed account of her present difficulties. I say to you after all this, "You really ought to pay Aunt Tillie a visit." I think you should have certain feelings, and that you should act on them.

According to the account of this case given by Brinton (212), it could be quite correct for the mother to say that the daughter, as recipient of the appeal above, ought to have such feelings. And her having such feelings could constitute a good reason for her visiting Aunt Tillie.

What this case shows is that appeal to feelings or emotions can be a legitimate basis for argumentation. The case's dialogue context is that of a persuasion dialogue, but it is not that of the critical discussion where one participant has the goal of proving a thesis, critically questioning a thesis, or trying to advance arguments to convince another participant that his thesis is true. Instead, the case of Aunt Tillie is an action-directed dialogue where the goal of the mother is to persuade the daughter to act—that is, to a specific course of action—on the basis of certain feelings that the mother is trying to raise in the daughter.

Why would an appeal to emotion be legitimate in this kind of action-oriented discussion? The reason in this case stems from the subject of the discussion, which concerns feelings of gratitude for past acts of kindness. Of course, the daughter has certain obligations to her aunt, as a family member and as a person who has been a recipient of her help. But her mother is also arguing that her daughter should have some feelings toward this person who has been very kind to her in the past. Given the nature of the discussion and the aims of the mother, which seem perfectly legitimate and appropriate, the appeal to pity would appear to be just the right means for the mother to use in trying to convince her daughter to act in a certain way. As goal-directed practical reasoning aimed at persuading her daughter to act, the mother's argumentation appears to be relevant and appropriate in the discussion.

Appeal to pity often seems inappropriate or questionable in political debates when used in place of hard evidence to plead for special interest groups or causes. However, in some cases, it is difficult to deny that an overt and highly emotional appeal to pity is worthy of attention and action.

In the following case an overt appeal to pity was made in a parliamentary debate (*Canada: House of Commons Debates*, 19 January 1987, 2357) to solicit funds to help Mr. Joey Smallwood, a well-known and respected figure in Canadian politics, finish a project he had undertaken in his old age.

Case 4.2

National Heritage
Aid Sought for Completion of Newfoundland
Encyclopaedia

Mr. John R. Rodriguez (Nickel Belt): Mr. Speaker, I am rising to urge the government to act quickly to help out Canada's only living

Father of Confederation, Joey Smallwood. Mr. Smallwood, now in his eighties, is being financially hounded owing to his failure to complete his beloved project, a Newfoundland and Labrador encyclopaedia, because he is not well.

I would suggest that the government pick up the cost of the completion of the valuable project started by Mr. Smallwood. This encyclopaedia is an invaluable resource not only for the people of Newfoundland and Labrador but for all Canadians and, as such, would be well worth the federal government's investment.

It is a great travesty to see one of Canada's living legends being forced to deal with sheriffs at his advanced age and in his poor state of health. Surely the government can find it in its heart to act now.

In the second paragraph this appeal addresses the question of whether the encyclopedia project is worth funding in order to see it completed. That question needs to be judged on the merits of the project itself. But on the presumption that the project is worth funding, the appeal to pity in drawing attention to Mr. Smallwood's difficult situation does not really seem out of place.

If the conclusion that the appeal to pity is supposed to support in this case is that the project is worthy of funding, the appeal to pity is not very helpful or relevant; one needs to look at the evidence on the merit of the project. But if the conclusion is that the project should be funded, then the public debt to Mr. Smallwood might be a legitimate reason for funding, especially on the presumption that the project is worth funding.

The appeal to pity is relevant in this case because the respondent is being called upon to help, or have sympathy for, a person who has helped the respondent in the past and who is presently in a difficult situation where he needs help.

Of course, the appeal to pity in this case is also part of a larger argument that depends on the worthiness of the project to be funded. As such, it could be a pivotal argument that could rightly be used to swing a burden of proof toward one side or the other in a case where a decision for action or inaction needs to be reached.

This case is a frank and overt appeal to pity that focuses on the person and situation of one individual who was loved and admired by many people in Canada. It is therefore a highly emotional and personal appeal, with patriotic *ad populum* elements mixed in as well. But despite the tradition of categorizing this kind of argument as a fallacious *argumentum ad misericordiam*, it would seem to be a mistake to automatically classify this case as a fallacy or paralogism.

This particular argument occurred in a segment of the House of Commons debates where short recommendations for financial initiatives or changes to legislative bills can be put forward. In this context, the appeal for funding for the encyclopedia project fits in well with the kinds of speeches that are appropriate. From a point of view of the rules of debate, this particular speech does not appear to be objectionable or out of place in any way. Of course, it is a short speech and leaves much unsaid about the worth of the project in question. But even so, it would contravene fair practices of the charitable interpretation of argumentative discourse to declare it a fallacious *argumentum ad misericordiam* just because it is such a heavily emotional appeal to pity.

It could be noted here that describing the *ad misericordiam* type of argument specifically as the appeal to pity, rather than as, say, the appeal to sympathy, or the appeal to emotional support for someone who needs help or sympathy, makes the argument seem dubious, lending credibility to the assumption that it is generally fallacious. If asked whether they want pity, disabled persons or returning war veterans will probably say "no," for 'pity' has connotations of condescension. To pity someone is to imply that the person pitied is in a pathetic state or situation and that one should be sorry for him or her. On the other hand, a returning veteran or a disabled person who is or has been in a hard or difficult situation and has suffered might appreciate sympathy, the understanding that someone else can have of the burden he or she has been carrying.

Calling the appropriate appeal to emotion in argumentation an appeal to *pity* therefore already stacks a weight of presumption against it, suggesting somehow that such an appeal is inappropriate, wrong, or fallacious. If it were called "appeal to sympathy" this connotation would be removed.

Charitable Appeals

Charitable appeals are quite interesting to consider in relation to the *argumentum ad misericordiam* because they are direct, often overtly pictorial, appeals to pity that can hardly be condemned as fallacious *per se*. A charitable appeal is a request for action, usually in the form of a financial contribution. Unfortunately, such appeals, especially the door-to-door or mass mailing variety, do sometimes turn out to be fraudulent or badly mismanaged. In other cases they are allied with special interest groups or political causes in a way that may not be made clear in their appeals. But this does not mean that the appeals to pity in these pleas are all misguided or misleading, much less that they are all cases of the *ad misericordiam* fallacy.

One has to look at the context of dialogue. A charitable appeal is an argument, but it does not normally seem to be a critical discussion of a disputed issue. Typically, it is an argument for action to solve or ameliorate an alleged problem. The appeal is premised on a need for action. Usually it focuses on the need of a particular group and may even cite an individual case that takes the form of an appeal to pity.

The case below was a full-page advertisement requesting readers to send money for the relief of famine victims in Ethiopia (*Newsweek*, 4 March 1985, 75). A photograph showed an emaciated child, crying and obviously in severe distress, juxtaposed to the headline: ETHIOPIA: THE MOST DEVASTATING HUMAN CRISIS OF OUR TIME. The text of the advertisement called for action to help with this crisis.

Case 4.3

THERE **IS** SOMETHING
YOU CAN DO ABOUT
THIS TRAGEDY . . .

You've seen the news reports . . .
* Thousands of people a day are starving to death!
* More than 6 million people are threatened by starvation.
* More than 100,000 could die from hunger and its related diseases in the next 60 days.

THE TIME FOR ACTION IS NOW!

HERE'S WHAT YOU CAN DO TO HELP!

Your gift of $15 is all it takes to feed a hungry child for a month! Just $30 can feed two children for a month. And $75 will provide emergency food for an entire family of five for a month!

The argumentation in this type of plea for action is based on practical reasoning. A need or problem is posed, and the action needed to solve the problem is recommended. Although an appeal to the emotion of pity is made, that in itself is not fallacious or unreasonable. Here, the emotional appeal appears to be quite appropriate. There is good evidence that many innocent people in Ethiopia were starving to death and desperately needed help. The reader's response should include the emotion of pity, and it should be a basis for action.

In fact, however, getting food to the starving people in Ethiopia turned out to be a lot more difficult than one might have thought. Roads were bad or nonexistent, and there were political problems in securing passage of the food. Thus, for those considering responding to an appeal for funds, practical questions should be raised concerning the ability of the agency sponsoring this advertisement to do the job.

The following charitable appeal was a mass mail-out letter sent out to solicit funds for the support of research, treatment, and support programs for hemophilia, the "bleeder's disease" caused by an absence of the substance that makes a person's blood clot. The appeal in the letter centers on the case of a young boy. His picture is enclosed. It is the face of a blond boy around nine years old, a face that is appealing except that it is severely bruised. He has two black eyes and he frowns sadly.

The letter says that this boy and his parents can handle the situation most of the time, thanks to research.

Case 4.4

> Most of the time, they can handle it. But sometimes—sometimes there is a desperate race to the hospital. Bleeding begins in Patrick's neck, the blood swells and starts to close the windpipe . . . a blow to the head starts internal bleeding in his brain or his spinal column. . . . These are the things that most often cause the death of boys like Pat, no matter how brave they are, no matter how careful they are, no matter how much they want to live. These are some of the reasons why we want desperately to find a permanent answer to the problem of hemophilia.

The letter goes on to say that it is "incredibly hard" for "any decent adult" to "stand by and do nothing while a child is suffering great pain." Following this statement a plea is made to support research on hemophilia, to supply blood products for hemophiliacs, to create hemophilia programs, and to send hemophiliac youngsters to a medically supervised summer camp. The letter closes by asking for a contribution to help this child and the others like him who suffer from hemophilia.

The centerpiece of the appeal is an enclosure with the letter featuring the photograph of the face of the little boy with his black eyes; under the photograph is a note from him in childlike handwriting, signed with his name.

Case 4.4a

> I got hit from inside my body by hemophilia which I got ever since I was borned and whenever I play hard or fall down or even for no good reason at all I can get hit again and again from the inside and end up looking like this on my face or tummy or all over and surprisingly find myself in the hospital with mom and dad all upset while they tell me that someday a cure will happen if enough people care but right now I have to be extra careful all the time and especially next time I play with georgie.

The letter closes with the statement that there "*must* come a day when we can free children from the danger and pain of hemophilia." It is signed by the "President Elect" of the National Hemophilia Society. Enclosed is a notice saying, "With the threat of another postal disruption it is urgent that you rush your donation back to us," giving instructions how to send it.

The use of the photograph and the note above is such a dramatic appeal to pity that perhaps one might be inclined to classify it, after Weddle, as a case of "milking the issue" or trying to push pity beyond its reasonable weight as a claim to consideration. But remember that the context of the dialogue is a plea to help those afflicted by hemophilia, whose victims clearly suffer and whose need is both urgent and deserving.

One problem brought out by this case is that the packet of appeal materials is too "self-contained," in the sense that the reader is typically not in a position to make appropriate judgments about the trustworthiness of the agency or the truth of the factual claims they have made.

Normally, the appeal to pity works best as an argument in a borderline case where the arguments *pro* and *contra* are balanced, so that the appeal to emotion can tilt the weight of presumption just enough to make a significant difference in the outcome. Appeals to emotion are typically weak arguments, but they can give an argument that added little push needed to make it swing to one side of a disputed issue.

However, in case 4.4 the appeal only fulfills this function if there is already a presumption in place that the argument as a whole is a good one. And typically, for the average respondent, this presumption would not be in place. Unless the recipient of the package already has a good reason to trust the credibility of this particular relief agency, she or he would likely throw this material in the garbage and send any charitable donation to some well-established relief agency known to be reliable. Here the appeal to pity may fail, not because such appeals are inherently fallacious, but because it lacks the right evidential backup to make it successful.

In a case like this, then, even if the appeal to pity appears melodramatic and heavy-handed, or deliberately, emotionally exploitive, still this does not mean that the appeal is necessarily a fallacious *argumentum ad misericordiam*. It is an argument that appeals to pity, but given that the appeal is to help those in an unfortunate position, the appeal to pity is a legitimate motivating consideration. In such a case, drawing attention to the misery requiring assistance could be a part of an appropriate argumentation for the situation, even if it is too self-contained to carry much weight by itself.

Of course, in the charitable appeals and other cases studied so far, it is possible for the appeal to pity to be weak, erroneous, inappropriate, or even fraudulent or fallacious. Much depends on the particulars of the given case. But in these types of cases, the appeal to pity is more than marginally relevant. It is the central basis of the argument and quite properly so. In this type of case, the burden of proof must be squarely on the critic who would make any allegation of the fallacy of *ad misericordiam* to back up any charge with solid evidence.

Excuses

Many appeals to pity are excuses. An excuse is a kind of speech act that is put forward by a proponent in a situation governed by a rule to which the proponent pleads exemption because of his special circumstances. The excuse is a plea for exception to the rule. The other party to the dialogue is the person whose job it is to apply the rule, and this respondent must judge whether the plea is acceptable or not.

Any teacher who has to grade assignments for credit is familiar with being a respondent in such cases. A typical example is the following case, similar to one discussed in Walton 1989 (102).

Case 4.5

> I know that this assignment is overdue, but I don't think you should deduct a grade, according to the rule for late essays stated in the course outline, because my grandmother died last week, and I have been feeling very bad about it.

In such cases some pleas are clearly acceptable excuses, for example, "Here is a note from my doctor, saying I was too sick to work on the assignment." Some are clearly not acceptable—"My goldfish died"—and others are borderline cases. The borderline cases can be dangerous for the respondent to admit too lightly. If allowed, they could set a precedent, encouraging other pleaders to cite similar excuses. One danger for the respondent is that of the slippery slope possibility, where so many exceptions become routinely allowed that the rule becomes useless or unenforceable and is overwhelmed by the exceptions.

The dialogue in such cases characteristically takes the form of a back-and-forth shifting of burden of proof. The student may argue, "You accepted Mary Smith's essay a week late, and my situation is the same as hers." This argument from analogy exerts pressure on the teacher to respond. In order to rebut the implication of being unfair or unreasonable (or bestowing "special favors"), the teacher must either give in, or show why Mary Smith's case was different.

In judging a case like 4.5 the person appealed to must look at the facts of the case and see whether they fit the rule and the established exceptions that have been allowed in interpreting the rule in the past. However, in a borderline case, good judgment is needed, and the respondent may need to look at the particulars of the case carefully. How badly did the death in the family affect this person, and would it have been unfair to expect completion of the assignment in that situation. A good response should be sympathetic, but fair.

In this type of case, rules and facts are the greater part of the serious evidence to be weighed in arriving at a decision. But where latitude exists, it could be a good thing to give the pleader the benefit of the doubt, if she or he appears to be sincere, and the situation merits real sympathy.

Generally speaking, the initial burden of proof is on the pleader to show why a case should be allowed as a legitimate exception to the rule. This is presuming that the deadline was clearly announced in advance and that nobody had objected to the deadline as unreasonable. Once the plea has been brought forward by the proponent, the burden of presumption shifts to the respondent to give a reason for her ruling in reply.

The kind of dialogue involved in this situation is not a critical discussion, nor is it exactly a negotiation. It is really a judicial hearing, involving the application of a set of rules to a particular case, often in an institutional setting with a decision-making structure of committees or authorities. Argument from analogy is an important part of the argumentation involved. The similarity of one case to another is a key consideration in judging the worth of an argument. The respondent needs to use, and to show, good judgment in deciding whether a plea is acceptable or not, judgment exhibited in the reasons given to back up her decision.

In case 4.5, the respondent will have to give reasons for not accepting the plea put forward by the pleader. She might, for example, cite cases of other students in the same class who have similar hardships, but managed to hand in the essay on time. Or she might elaborate on the rule and various precedents, arguing that such-and-such a type of excuse is normally not judged sufficient to create an exception.

The fact that such a plea appeals to pity does not automatically rule it out as an acceptable plea, much less make it a fallacious argument. However, appeals to pity are often used to bolster a clearly inadequate excuse. And so, one rightly tends to be somewhat suspicious of a particularly colorful or exaggerated appeal to pity. In some cases—look ahead to case 8.1 for an example—the appeal to pity becomes fallacious when it is pressed ahead too hard and used as a device to shift to a different type of dialogue. But appealing to pity is not, in itself, irrelevant; nor is it evidence of mischief or fallaciousness.

Hart and Honore (1969) have pointed out how ascriptions of criminal responsibility in law are defeasible by appeal to excusing considerations. For example, a defendant can plead that he did carry out an illegal act but argue that he did not do it voluntarily, because he was coerced or did not know what he was doing at the time. According to Hart and Honore, then, allegations of responsibility are defeasible, meaning that they are subject to defeat in exceptional cases. Defeasibility, therefore, seems to be an important factor in evaluating excuses, although appeal to pity is not one of these recognized categories of excuse for defeating responsibility in law.

It seems then that in the context of excuses, appeal to pity in the argumentation is not altogether irrelevant. Or, at any rate, it would be difficult to exclude appeal to pity as entirely irrelevant. In explaining why a particular deadline was not met, for example, it could be quite legitimate for a pleader to appeal to pity in describing the particular situation that made the deadline unreasonable. Such an appeal can have some argumentative value in shifting a small weight of presumption to one side, and where the evidence is balanced on both sides, the decision could go either way.

In the context of argumentation concerning excuses, appeal to pity is not a totally irrelevant or fallacious move *per se*. But it is a defeasible move—a move that can legitimately be put forward for consideration, and equally legitimately be rejected or rebutted as a good argument for making an exception to a rule.

In this type of argumentation, however, the appeal to pity would seem to be, by itself, a weak consideration that serves best when it is supported by other arguments, including factual considerations relevant to the case in point. They are not necessarily irrelevant or fallacious, but they can easily be blown out of proportion by their emotional impact. The real problem, most often then, is how to counter such an appeal by putting it in a proper perspective of its context of dialogue. What is the best way to react to an *argumentum ad misericordiam?*

Countering Relevant Appeals

Although *ad misericordiam* arguments often are treated by the textbooks as fallacies because they are irrelevant, in many cases they are relevant to the issue of a critical discussion. The problem in such cases is seeing how the appeal to pity is relevant.

Some arguments that appeal to pity can be very persuasive and can be used powerfully to refute an argument, even if their real logical weight as an argument is less than it seems. Emotional arguments that can have a powerful impact on a particular audience, for example, may be used aggressively to seize control of an argument.

When this happens, it can be difficult to prevent oneself from surrendering to the emotional appeal. To evaluate argument effectively requires that critical detachment not be swayed by an emotional appeal of the moment.

Consider a case where two people, Helen and Bob, are engaging in a critical discussion on the subject of tipping. Bob is arguing for the thesis that tipping is a good practice that should be maintained. Helen is arguing for the thesis that tipping is a bad practice that should be discontinued. Suppose that in the midst of their discussion Bob presents the following argument.

Case 4.6

> Many people who are barely surviving on low incomes depend on tipping to support their families. Let me introduce you to Arlene. Arlene is a single mother of four small children. She depends on the tips she receives from her job as a waitress to support her little children, Jeremy, Arabella, Bertram, and little Debbie. Look at poor little Debbie. She has to dress in rags because her mother cannot afford to buy a dress for her [at this point, Debbie starts to cry, and the rest of the family tries to comfort her]. Now as I understand it, Helen, you are advocating trampling on the rights of unfortunate low-income families by choking off their source of support. How can you be so heartless as to argue for this drastic measure?

In presenting the argument above, Bob has appealed to pity. In traditional logic the question would be raised whether he has committed the *ad misericordiam* fallacy. Bob may be making a worthwhile point in his argument, that eliminating tipping could place low-income persons into greater hardship. But one could argue that his dramatic way of doing it, as above, is an emotional appeal calculated to make Helen's argument look so bad, so unfair and reprehensible, that an audience would be unduly swayed and diverted by it and would reject Helen's side of the argument altogether.

How should Helen react, ideally, in the context of the critical discussion on tipping? She could accuse Bob's argument of being irrelevant to the real issue of the discussion. But it is relevant, as Bob could show: if discontinuation of tipping would increase hardship for many low-income families, then this factor should be taken into account in judging whether tipping is, on balance, a good or a bad practice.

Bob's argument can be shown to be relevant using the implied argumentation scheme for practical reasoning. Bob argues that tipping is a necessary condition for the well-being of low-income families, that discontinuation of tipping would create a hardship for these families. Therefore, he argues, one should not be against tipping.

In principle, there is nothing wrong with appealing to pity or sympathy in an argument, and such an appeal cannot be taken in every case to be a fallacy. But if the appeal to pity diverts or distracts the flow of dialogue from other relevant lines of argument by striving for a highly dramatic impact, there can be a problem. In this case Bob follows up his dramatic appeal to pity by describing Helen's position in exaggerated language, claiming that she advocates "trampling on the rights of unfortunate low-income families" and "choking off their source of support." In conjunction with the appeal to pity, Bob's loaded insinuation makes Helen's side of the argument look so bad that she will have great difficulty replying. To survive in the argument, she will have to rebut these charges. And it will be difficult for her to do this effectively without making an equally dramatic show of emotion—perhaps righteous anger or outrage at being accused of trampling on poor unfortunates.

Helen should react by objecting to Bob's loaded language of "trampling on the rights" of people and "choking off" their means of support. And she should also object to the allegation that she is "heartless" in arguing for this "drastic measure" of discontinuing tipping. To back up these arguments and reply more positively, Helen could go more deeply into the causes of the pitiable situation described by Bob.

Helen could argue that it is the institution of tipping that is one of the causes of the plight of low-income families. She could argue that if these persons in service industries were paid a decent basic wage, they would not have to rely on tipping to buy their necessities of life. Thus Helen could challenge the practical reasoning behind Bob's argumentation in his appeal to pity. She could concede that the plight of this low-income family is a genuine problem, saying that she is as concerned about it as Bob is. But she needs to argue that Bob's proposed solution to the problem is not the right approach. She needs to argue that it is tipping-dependency that leads to the pitiable situation of this family and that the abolition of tipping will enable them to be paid regular wages, making them less dependent on the fluctuating economy or the whims of their customers.

Here the best response to the appeal to pity is not to reject it as irrelevant or fallacious but to acknowledge its relevance to the issue and turn the argument on its head. The best and most effective tactic is to turn the argument against the other side.

What must be noticed, however, is that it is not always so easy to counter a relevant appeal to pity. The use of a pictorial appeal to pity can have such a powerful emotional impact that responding to it may require lengthy assurances and recommendations for dealing with the particular case cited. Thus the use of a powerful pictorial appeal to pity can pose a single issue that comes to dominate the critical discussion.

Case 4.7

> A story on the news program "20-20" discussed the current controversy on the ethical question of wearing fur garments. On the *pro* side, advocates for the furriers argued that animals in the wild often suffer cruelties of nature more severe than human trapping and that animals raised in captivity for their furs are comparatively well treated. On the *con* side, film clips were shown of animals in leg-traps bleeding profusely and obviously in severe distress. One clip showed a fox in captivity being killed. It had metal pincers fastening its jaws shut, and a metal rod shoved into its anus closed an electrical charge that electrocuted the shuddering fox.[3]

The person watching this program with the author started to cry out and jump around as these pictures came on the screen. She was clearly upset by this visual presentation, even declaring at the end of the program that she would never buy a fur coat.

For the *pro* side to counter this appeal to pity would be a formidable task. It would require all kinds of reassurances that cruel practices of the sort portrayed can be eliminated. Even so, it might be hard to neutralize the impact of such a gruesome picture. The impact can be overwhelming.

This case can be contrasted with some of the previous ones, where the appeal to pity was a weak but relevant argument that just gave enough of a tug in one direction to tilt a balanced set of presumptions; in case 4.7 the appeal to pity is so strong as an immediately present argument that it seems to blow everything else aside. In this case the psychological response elicited by a graphic representation of a repellent action overwhelms the balance or "distancing" needed to critically question the argument, that is, to put the graphic representation in a critical perspective. This problem is analyzed in Chapter 8 in relation to the functioning of critical doubt in argumentation.

---

3. This case is presented from the author's memory of viewing the program; it is not a *verbatim* transcript.

The Case of the Non-Smokers' Health Act

The *argumentum ad misericordiam* in the following case is a species of argument from consequences, citing the bad effects of a proposed parliamentary bill on a particular group of people as a reason for not supporting the bill. The proposed legislation in question was Bill C-204, an act to regulate smoking in the workplace and to restrict cigarette advertising. Speaking against this bill (*Canada: House of Commons Debates*, 12 February 1987, 3390–3391), the Honorable Bud Bradley outlined the plight of the proud but struggling tobacco farmers who would be "downgraded" and become "outcasts" if this bill were passed.

Case 4.8

Private Members' Business
—Public Bills

[English]

Non-Smokers' Health Act
Measure to Enact

[3390]

The House resumed from January 23, consideration of the motion of Ms. McDonald (Broadview–Greenwood) that Bill C-204, an Act to regulate smoking in the federal workplace and on common carriers and to amend the Hazardous Products Act in relation to cigarette advertising, be read the second time and referred to a legislative committee.

Mr. Bud Bradley (Parliamentary Secretary to Minister of National Defence): Mr. Speaker, I appreciate the opportunity to speak today in this legislation. The whole thrust [3391] of this legislation, which is supported by the New Democratic Party, the Liberal Party, cancer societies and anti-smoking groups, is to create a smoke-free Canada by legislating smoking out of Canadian life. But have we forgotten the Charter of Rights, the freedom of choice, the right of the corner store to survive, where some 70 per cent of its income is from confections and cigarettes? Have we forgotten the rights of people to prosper, to work, to raise a family and to be proud? These are the rights of a group of people I would like to address today, the

Canadian tobacco farmers. These are people who over generations have developed blow sand into soil and developed a healthy plant from a weed. In most cases they were immigrants who became the proudest of Canadians. No one faulted them back then. They received nothing but praise for their contribution to agriculture and to their country. Tobacco turned into a valuable commodity. Billions of dollars were made and billions of dollars were taxed but those billions of dollars were not for the farmer. Families from all over Canada came to work. Students paid their way through university working in the fields. Cigarette manufacturers developed empires. Wholesalers built businesses. Corner stores flourished, and vending machine operators haven't done badly either.

The industry employs some 62,000 people across Canada. Consumer spending provides over $6 billion to the Canadian economy. Federal and provincial governments collect over $4 billion a year in taxes. But what about the farmer? He receives six cents out of a $3 package of cigarettes. We have now ascertained that smoking is a health risk. I do not think that finding is questioned any longer in Canada. However, we now have a group in society, once admired for their hard work, dedication, ingenuity, and perseverance, who have all of a sudden become outcasts. No one thought of this problem back then. Farms that at one time were worth $500,000 to $1 million cannot be sold. Their agricultural peers are downgrading these farmers. Society is shunning them. Governments are taxing and legislating them out of business.

In the tobacco growing areas of Canada we have bankruptcies, family breakdowns and suicides. What about those farmers? What about the 2,600 dedicated tobacco farmers in Canada? What does the legislation do for them?

The Minister of Agriculture (Mr. Wise) is attempting to help. If it had not been for his intervention, the 1985 crop would not have been sold. He provided $90 million in advance crop payments last year, many times what had been provided in the past, and he will be providing $65 million this year. The capital gains exemption, fuel tax rebate and rural transition program are welcome moves.

The Minister of Finance (Mr. Wilson) provided taxation to fund a REDUX program, some $30 million in last year's Budget. That was the first time the tobacco farmers' plight was ever recognized in a Budget Speech.

The Minister of Agriculture is announcing today the establishment of an alternate enterprise initiative program to assist diversification initiatives in the tobacco growing regions of Canada. It is a program to develop a co-operative approach towards increasing the production and marketing of alternate crops without causing significant disruption to existing Canadian primary production. This $15 million program will focus on the development of new crops, production technologies, marketing and processing opportunities. It is the first part of the $30 million tobacco diversification plan announced in the 1986 Budget. The results of discussions with the industry on the second part should be announced in the near future. I applaud this very positive step developed through co-operation at the federal grower level.

It is obvious that the inevitable is happening. The demand for this product is shrinking fast. Tobacco farmers have been told that their product is not wanted in Canada, but they are also being told, "We do not want you to export health risks to Third World developing nations either." Eighty-six per cent of Canadian tobacco exports go to developed nations headed by the United States, the United Kingdom and Germany. It is hard for the growers to understand why Canada will not support the export of a crop which some other country will then provide.

Alternate crops are being looked for, something to reduce the growers' dependency on tobacco. But what can one grow on 100 acres which will pay a $500,000 mortgage? At the present time much work is being done to research alternate crops, alternate crop processing, and alternate crop marketing. However, time and money are needed and these are two commodities of which they are short. Even with the assistance provided to date by the Government, more than was ever provided in the past, some 500 to 700 farmers will be forced out of tobacco production this year.

I offer, Mr. Speaker, first, that as long as Canadians smoke, let Canadian tobacco farmers have the right to grow the tobacco. Second, as long as we, society, the provinces, and the federal government, tax and legislate tobacco farmers out of business, we have an obligation to assist them. If my colleague the Minister of Finance does anything with the special tax levied on cigarettes last year to assist these people, I hope he uses it to increase the funds available. Let us provide the tobacco farmers the needed time and money to alter their direction. Sure, they demonstrated and protested. They want to survive. They want the right to do what they trained for and what they do well. I ask my colleagues if they would not fight for their business, home, family, and future if the whole country seemed to be against them? I urge my colleagues on both sides of the House to consider this one comment; do not turn your backs on this group of proud Canadians.[4]

Mr. Bradley appeals to pity when he describes the tobacco workers as people "who over generations have developed blow sand into soil" and who were immigrants in most cases. These dedicated and persevering people who have worked so hard have "all of a sudden" become "outcasts" by the government "taxing and legislating them out of business." Mr. Bradley even goes so far as to cite "bankruptcies, family breakdowns and suicides" in the tobacco growing areas of Canada.

This last point in particular is somewhat weak because Mr. Bradley has not cited evidence to show that these social problems are caused by legislation on tobacco growing; the same problems can be found in many other areas of Canada as well.

Although Mr. Bradley's argument is weak—and it would be consistent with the standard treatment to dismiss it as a fallacious *argumentum ad misericordiam*—it is a reasonable form of argumentation. Basically, his argumentation follows the scheme for the argument from consequences, as follows: if the consequences of a line of action A are bad, then one should avoid the line of action A. According to Mr. Bradley, if Bill C-204 is passed, bad consequences will befall the tobacco farmers of Canada, and therefore this bill should not be passed. This in itself is a reasonable argument, but it has to be evaluated in the context of the larger sequence of practical reasoning of which it is a part.

Mr. Bradley's speech dealt with the rights of the tobacco farmers and the impact of Bill C-204 on their means of making a living. However, this is only one aspect of the bill that needs to be considered. The purpose of the bill is to protect the health of all Canadian citizens and to clarify the rights of nonsmokers. This larger picture was brought out effectively by the reply made by the Honorable Dan Heap (3392).

4. I would like to thank Gene Benoit for drawing my attention to this case.

Case 4.8a

Mr. Heap: The Hon. Member spoke about the rights in terms of the Charter. He does not give any clear direction as to what he wants to do with respect to the rights of people who are being adversely affected by the tobacco industry and the use of its products. He has suggested that there is an absolute right to grow, manufacture, sell and smoke tobacco. In fact, he seems to claim that that is a Charter right. I question his reasoning on that point. I do not think there is any such absolute right. There are very few absolute rights.

If we are to talk about rights I say that the right to health is much more fundamental than the right to be a tobacco farmer, a Member of Parliament, or the driver of an automobile which emits lead into the air which pollutes our cities. There are many activities which can be challenged for their effect on other people. The right to pursue them cannot be considered absolute without considering their effects on other people.

The whole reason for the campaign in the matter of smoking is the effect on other people. We are told about the number of people and the number of dollars reflected on the part of growers, manufacturers and retailers. But what has to be weighted against that is the fact that tobacco is associated with more deaths and illnesses than any other single product in our country. Some 35,000 people in Canada die every year from tobacco-related diseases. That is more than the number of people who produce tobacco. I hope the Hon. Member does not seriously argue that 26,000 farmers should be allowed to contribute to the deaths of 35,000 people. By making such an absolute claim to rights that is the logic he is putting forward.

This argument points to the positive consequences that may be expected to follow from passing Bill C-204—or, to put it in the form of a double negative, the negative consequences of not passing it. It is therefore a counterbalance argument to Mr. Bradley's negative argument from consequences. Mr. Heap's argument has the form of a positive (opposed) argument from consequences: if the consequences of a line of action *A* are good, then one should implement this line of action. According to Mr. Heap, if Bill C-204 is passed, it will stop many people from being badly affected by tobacco-related health problems and thereby also save the

Canadian people a lot of money every year in costs of health care. Not only that, it will save many lives every year. This is a positive consequence. According to Mr. Heap's counterargument, these positive effects of Bill C-204 outweigh the negative effects on the tobacco farmers. The one argument is opposed to (a negation of the point of view of) the other.

This rejoinder shows the bias in Mr. Bradley's argument—an argument that considers one side of the issue only and overlooks factors on the other side that would be fairly obvious to most respondents—and therefore may explain, to some extent, why it appeared at the outset to be a fallacious *argumentum ad misericordiam*. Note however that Mr. Bradley's argument should not be rejected as worthless. The welfare of tobacco workers is an appropriate, even an important, matter for discussion in the House of Commons. What action should be taken if farmers are forced out of tobacco production, as Mr. Bradley alleges, is a legitimate issue for discussion in relation to the debate on Bill C-204. Citing negative consequences of this bill on the employment of tobacco workers is, in principle, a reasonable line of argumentation.

What is objectionable about Mr. Bradley's argument, however, is its one-sided nature. It does not tell the whole story. By appealing to pity, it attempts (not very effectively) to make a weak, rebuttable argument appear stronger than it really is. This tactic certainly bears many of the hallmarks of the questionable *argumentum ad misericordiam*.

When deliberating the practical wisdom of a legislative bill, it is reasonable to take into account not only the presumed goals of the bill but also its significant positive and negative side effects, to the extent that these can be reasonably conjectured. While negative side effects can be a good reason for not going ahead with the proposed action, this factor alone is not normally sufficient to rule out the action as one that is prudentially unreasonable. One has to take into account other factors as well, including the goals of the action, to the extent these are known, and the positive effects of implementing these goals by the action, to the extent these can be estimated. In deliberating a course of action, other factors like available alternatives and potential obstacles are relevant. The argument from consequences can be an important part of such a network of practical reasoning. And no argument of this sort should be regarded as complete until a serious attempt has been made to look at both sides of the issue—especially including the weighing of the positive versus the negative argument from consequences.

Pushing a Questionable Presumption

In some cases an appeal to pity can mask the lack of evidence for a key presumption, using emotional impact to disguise the weakness of an argument. The following case concerns the use of a sonogram videotape picturing the suction abortion of a twelve-week-old fetus as a visual aid for antiabortion (pro-life) arguments. The fetus's movements on the videotape presentation strongly suggest that a baby is squirming to escape destruction by pincers tearing it apart. Needless to say, the emotional impact of viewing the videotape, called *The Silent Scream*, is very forceful.

The following argument on abortion was brought forward in *Canada: The House of Commons Debates* (5 March 1985, 2738f.).

Case 4.9

Abortion
Videotape of Abortion Process

Mr. Gordon Taylor (Bow River): Mr. Speaker, if a nuclear bomb were dropped on Ottawa probably 50,000 humans would die, and every newspaper and every government in the world would call it a disaster. Yet, every year in Canada, some 50,000 human babies are murdered and it is barely newsworthy. Perhaps it is because we do not really realize what happens when a baby is aborted.

Modern technology has enabled Dr. Bernard Nathanson to use a sonogram videotape to picture a 12-week-old child in the womb being aborted. A hollow plastic tube with a sharp tip is inserted and, using suction from a vacuum, the feotus [sic] starts to be torn apart. The baby retreats helplessly to the far side to escape the device, but it pulls off her legs and disembowels her. She struggles, her head falls back, and her mouth opens in agony. Then a pair of forceps crushes the brain, and another child is murdered.

The sonogram proves beyond a shadow of a doubt that the feotus is alive. It is a human being. He or she cringes and tries to escape death. We can watch and see the gruesome picture of a child being brutally murdered. The mouth opens in a cry for help. Can't you hear that silent scream?

This videotape, and even its description quoted above, conveys a powerful visual appeal to pity. To evaluate it, one should begin by asking what the conclusion is it is brought forward to support.

The issue between the pro-life and pro-choice arguments, in the context of the argument above, is whether a fetus is a human being (a person). According to Mr. Taylor's argument above, the sonogram proves "beyond a shadow of a doubt" that the fetus is a human being. But does Mr. Taylor's argument *presume*, before it *proves*, that the fetus portrayed in the videotape is a human being? Notice that he describes the fetus in the videotape as a "12-week-old child." Later he says that "the baby retreats helplessly," "her mouth opens in agony," and "another child is murdered." This language presumes that the fetus pictured in the videotape can be described as a person.

This use of definitions to support one side of the argument would, of course, be unacceptable to pro-choice exponents, or possibly to anyone who has not already accepted Mr. Taylor's conclusion that the fetus is a human being. Thus, one may easily suspect that Mr. Taylor's presumptive use of the language of "human beings" is a begging of the question at issue, an assumption of the thesis to be proved that marks his failure to prove it. For the present purposes it is enough to observe that Mr. Taylor's argument is weak because it uses language and assumptions that its opponents would not accept. However, by its strong visual appeal to pity, the argument may be so powerful its questionable presumptions may be overlooked in any partisan description of what is portrayed.

In case 4.9 evidence should be given to support its presumption that a twelve-week-old fetus is a human being. The two sides need to agree on the general criteria for defining a human being, and then the arguer has to show why a twelve-week-old fetus meets these criteria. Understandably, this is a hard burden of proof, but perhaps it could be met by careful arguments in the context of the dispute.

The evidence Mr. Taylor has given is that the twelve-week-old fetus pictured on the videotape *appears* to be exhibiting behavior consistent with or similar to familiar human actions—retreating, screaming in agony, etc. However, critics have alleged that this appearance is misleading, that the movements pictured in the videotape—suggesting attempts to escape, screaming, and so forth—are random. Who is right? It is hard to say. Perhaps medical evidence could better interpret these movements, or perhaps it is more of a philosophical question. However, there is more evidence to be considered before accepting Mr. Taylor's presumption that the picture must be interpreted as that of a child being brutally murdered. By taking the contested point as a presumption, Mr. Taylor is trying to avoid the need to fulfill the proper burden of proof needed to make his case convincing.

So Mr. Taylor's conclusion may be right, or it may not be, but his argument is phrased in a presumptively loaded way that tries to leave the person to whom the argument is directed no choice in how to interpret what he or she sees. It is a weak and questionable argument that tries to pass itself off as proving its conclusion "beyond a shadow of a doubt." This strategy would perhaps not work very well except that the picture of what appears to be a baby in agony being destroyed is so emotionally overpowering that the viewer or reader is shocked and traumatized. The viewer or reader cannot help henceforth associating abortion with strong emotional feelings of pity for the "silent scream." This appeal to pity is so powerful in its appeal to deep emotions that it is easy to overlook the basic weakness of the argument in relation to its strongly phrased conclusion.

The power of pictures to convey a persuasive appeal to pity in argument is well illustrated by the depiction of cuddly baby seals being killed with baseball bats. The depiction was part of the recent campaign to stop the seal hunt in Canada. This use of media material conveyed a powerful appeal to emotion that was highly successful and led to a ban on seal hunting.

A sequel to *The Silent Scream*, called *Eclipse of Reason*, does not purport to exhibit the reactions of the fetus. Instead, as described in a *Newsweek* article, it uses a camera positioned at the mother's feet along with an intrauterine camera scope to show bloody fetal material being pulled from a woman's vagina during an abortion (Gelman and Miller 1987, 32). According to the *Newsweek* report, this videotape is even more graphic and bloody than *The Silent Scream* and is "harder for critics to dismiss as misleading." Because of its gory content, the film had a strong impact on viewers.

However, it is hard to say in this case how the emotions aroused by the film might move those who are undecided about the abortion issue or provoke those who are already committed to one side or the other to change their minds. The danger of this appeal to emotional reaction is that it may divide the extreme positions even more sharply and distract from a more reasoned examination of the arguments at issue.

Once again, an appeal to pity is in itself not necessarily fallacious in the context of a critical discussion on the issue of abortion. It is how the appeal to pity is used that is problematic. Pity is such a powerful emotion that it can conceal the weakness of questionable presumptions.

The pictorial appeal to pity can be quite legitimate, in principle, in some cases of critical discussion. But even if it is legitimate in what it does, it may always be wise to remember to ask what is not done. In approaching any appeal to pity in an argument, a first step is to determine what the real conclusion or conclusions of the argument are, or should be. Whether the appeal in the argument is strong or weak depends on what the conclusion is and how presumptions are brought forward and supported (or not) by the required evidence.

Appeals to Pity and Loaded Questions

In addition to premises and conclusions of arguments, appeals to emotion can also be used in questions. Case 4.9 was of this type, but sometimes an appeal to pity can be used in other ways to disguise the weakness of a question.

The appeal to pity can be used to ask an inappropriate question that may, despite its inappropriateness, convey information or allegations that may make the answerer seem guilty or responsible. What should be noticed here is that the appeal to pity is being used in conjunction with the asking of a loaded question. The question is *loaded* in the sense that any attempt by the respondent to give a direct answer will automatically set a presumption in place that the respondent is guilty or is committed to something not in his or her favor. In the following debate, which took place in the Oral Question Period of the House of Commons Debates (vol. 128, 8 April 1986, 12000), Mr. Ian Waddell asked a loaded question, similar in some respects to the famous spouse-beating question. It is based on an appeal to pity, and it also has some elements of an *ad populum* appeal.

Case 4.10

Housing
Eviction of Roomers from Vancouver Hotels

Mr. Ian Waddell (Vancouver-Kingsway): Mr. Speaker, I have a question for the Minister of Labour who is responsible for housing.

The coroner for the City of Vancouver, to whom I spoke this morning, told me that on April 4 at 9:15 a.m., at the corner of Pender and Carroll Streets in Vancouver, Harold Scarrow, age 61, who was to be evicted from a room for which he was paying $225 a month at the Lotus Hotel, to make way for Expo guests paying $50 per day, threw himself under a dump truck and died. What is the Government of Canada doing to stop this inhumane treatment of poor people who cannot really defend themselves?

Hon. Bill McKnight (Minister of Labour): Mr. Speaker, I can inform the Hon. Member that with the changes in the National Housing Program which we are now negotiating, we will meet the needs of those needy Canadians like the one he has described twice as fast as under the old program. I can inform the Hon. Member that last year the expenditures on social housing in the Province of British Columbia—

Mr. Turner (Vancouver Quadra): Went down.

Mr. McKnight: —increased 23 per cent in 1985 over 1984.

Ms. Copps: Because it was the lowest in the country in 1984.

Request for Federal Intervention

Mr. Ian Waddell (Vancouver-Kingsway): Mr. Speaker, I do not know if the Minister heard my question. Given that at the end of February there was another suicide of a man evicted from the Patricia Hotel in order to make way for Expo guests, would the Minister—

Mr. Hnatyshyn: There's no Minister of Hotels.

Mr. Waddell: Would the Minister use the good offices of the federal government, through the Prime Minister, to deal directly with Premier Bennett of British Columbia and make sure that this situation does not continue?

Hon. Bill McKnight (Minister of Labour): Mr. Speaker, we recognize the concern of the Hon. Member and others in this matter. It does not fall within the jurisdiction of the federal government. The federal government is attempting to negotiate a housing agreement across Canada that will change—

Mr. Waddell: How many more suicides are we going to have?

Mr. McKnight: The Hon. Member asks how many more suicides there will be. I deeply regret the loss of life, but the Hon. Member must recognize that we are attempting to change a program, which under the previous administration only targeted one-third of its dollars to those in need, to a program that targets 100 per cent of the federal government expenditures to those Canadians truly in need.

Mr. Waddell's appeal to sympathy is for the plight of a man who was evicted from his room because of the rising accommodation rates occasioned by Expo '86 in Vancouver. According to Mr. Waddell, this man threw himself under a dump truck after being evicted.

This appeal to sympathy for an individual caught in a bad situation is not in itself unreasonable, but notice the form of Mr. Waddell's question following up his appeal. By asking what the government is doing to stop this "inhumane treatment of poor people who cannot really defend themselves," the question seems to imply that the government has condoned the inhumane treatment of defenseless poor people and is continuing to do so.

Mr. McKnight replied by referring to the National Housing Program, a federal government program. Mr. Waddell should have then questioned how this program was relevant to a situation like that of Mr. Scarrow, the man described in Mr. Waddell's cited case. However, instead of doing this, he once more launched into the attack by describing the suicide of another man evicted from a hotel in Vancouver.

Next, Mr. Hnatyshyn did Mr. Waddell's work for him by saying that there is no Minister of Hotels. In other words, he was questioning whether the cases cited by Mr. Waddell come under federal jurisdiction, for it seemed likely that the problem was really one that should be best dealt with by the City of Vancouver. Thus, the issue was raised whether Mr. Waddell's original question was appropriate to be asked in the federal parliament. However, Mr. Waddell made the connection in his next question by asking if the federal government could deal with the Premier of British Columbia to "make sure that this situation does not continue." In reply, Mr. McKnight firmly stated that the question does not fall within federal jurisdiction. In short, there is some reason to doubt that Mr. Waddell's question about the case of Mr. Scarrow was an appropriate one to be asked in the federal parliament.

Rather than leave the issue at this point, however, Mr. Waddell launched into a third attack with yet another aggressive question: "How many more suicides are we going to have?" In reply, Mr. McKnight simply repeated the question, indicating to all that the question did not need to be taken seriously as a sincere query. This third question was simply a continuation of Mr. Waddell's strategy of trying to fix blame for suicides on the government by asking aggressive questions that suggest that government policies are responsible for inhumane treatment of poor people who are, as a result, committing suicide.

At the end, all Mr. McKnight could reply to this onslaught of loaded questions was that the present government policy is better than that of the previous administration in spending funds on those in need. This was an attempt to bounce the ball back into the questioner's court. But it was perhaps not unreasonable, given the persistent and aggressive use of loaded questions by Mr. Waddell's *ad misericordiam* attack.

The problem with Mr. Waddell's questioning here is whether it is appropriately addressed to the federal government. To his credit, Mr. Waddell does try to establish a connection, but the appropriateness of his questioning on this issue, raising these particular cases, remains open to challenge. However, by basing his argument on a powerful appeal to pity and by weaving his appeal into highly loaded and aggressive questions, Mr. Waddell is trying to push ahead and discredit the government despite the marginal appropriateness of his question in this forum. The appeal to pity is an aggressive ploy that covers up the weakness of a dubiously appropriate line of questioning while using emotional impact to impute blame.

In itself, Mr. Waddell's question is not unreasonable. It is appropriate for him to ask the government minister what the government is doing to help poor people. In some respects it is quite commendable for Mr. Waddell to take an interest in the case of the unfortunate Mr. Scarrow.

What is questionable is the linkage between the government policy on housing and the unfortunate fate of Mr. Scarrow, as if there were a direct causal relationship between one and the other. If Mr. Scarrow's case were to be investigated in detail, it might turn out that there were many causes for his death. To blame the government for it in Question Period without giving the government minister a chance to look into the case before responding is to push a presumption of guilt very aggressively.

The charge of fallacy comes in not only in Mr. Waddell's loaded question at the end of his first speech. It comes in during the subsequent dialogue where Mr. Waddell's follow-up questions and attacks do not give enough room for his respondent to reply properly. To pin down the allegation of *ad misericordiam* fallacy, a critic must look at the whole text of dialogue as a connected sequence. Mr. Waddell tries to fix blame for the suicides by continually pressing ahead with the presumption that the government is responsible for inhumane treatment of poor people, by pressing ahead without acknowledging his respondent's replies or giving him a chance to challenge this heavy attack.

Evaluating Cases

In evaluating any allegation that an *ad misericordiam* fallacy has been committed, it is necessary to carefully examine the evidence that is available. At first sight, this approach may seem like an overly relativistic approach to the *ad misericordiam* that makes everything stand or fall on the details of the particular case, but it is not. It is even possible to discuss and evaluate general cases, where the exact or complete details of the text of discourse are not given or known. The point is that any evaluation of whether an *ad misericordiam* fallacy has been committed must be conditional, relative to what is known or given in a particular case.

In some cases, it may not even be a named individual who is alleged to have committed the *ad misericordiam* fallacy. It might be a group, like a corporation or a political interest group, whose arguments, campaigns, or rhetorical tactics are criticized for using questionable appeals to pity. Such a criticism could be quite reasonable and persuasive, in a conditional way, as an argument to shift a weight of presumption, even if a specific text of discourse is not quoted or cited exactly in the presentation of the criticism.

The following case is an essay written in *Newsweek* arguing against the stereotypical appeal to pity on behalf of the elderly as a sedentary, decrepit, and poor group who are especially needy, vulnerable, and deserving of support and tax relief. Samuelson alleges that this appeal to pity is based on a double standard, for as older people have been placed in a much better position in recent times, with much better health and social security benefits, they have also become much more energetic in defending their group rights and interests. They now argue that age doesn't rob them of vitality. But on the other hand—hypocritically, according to Samuelson—

they also like to depend on the old argument that age entitles them to special treatment because of their infirmities.

Case 4.11

> Flipping through *Modern Maturity*—the magazine of the American Association of Retired Persons—you find people who are energetic and adventurous, although not always retired. There's an item on a 68-year-old rock-music critic for a major newspaper. A travel article celebrates the glories of scuba diving. Moving along, you're treated to a dazzling selection of readers' photographs.
>
> Now consider the AARP as a political lobby. It's crusading to persuade presidential candidates to support more aid for long-term care: home assistance and nursing homes. The campaign's emotional power is enormous. One advertisement shows an elderly woman bathing and feeding her husband, who has been incapacitated by stroke. "I would say seven-eighths of the day and the night are taken up with him," she says. "I feel like a bird in a cage." (Samuelson 1988, 68)

The charge made by Samuelson is that the elderly like to rely on the *ad misericordiam* argument in their rhetoric, because it is still a very persuasive argument when posed as a powerful emotional appeal to pity. But, he asks, are they really entitled to use this argument when federal spending on medicare and long-term care for the elderly is now enormous? While the old may have lived in poverty in the past, Samuelson argues that with present social security and federal retirement benefits, the older population is now both wealthier and better supported through payroll taxes.

Samuelson criticizes the stereotypical appeal to pity for needy elderly people on two grounds. First, he argues that it is not really based on the current facts. Second, he argues that it is contradictory or hypocritical because it is inconsistent with their argument that they are vigorous, energetic, and capable of being independent. This second criticism is particularly interesting because it is a type of circumstantial *ad hominem* argument against the double-sided rhetoric of hypocritical, self-interested argumentation.

Samuelson doesn't accuse the AARP of committing a fallacy, or even use the term "fallacy" at all. But he could have used this term with some justification because the AARP rhetoric is not only not supportable by the preponderance of evidence but is contrary to commitments in their own position. Even though extensive texts of AARP speeches are not cited, the criticism carries enough weight of presumption to shift a burden of proof as a conditional criticism of fallaciousness.

The problem with many cases of appeals to pity is that the author does not give enough information to enable a reader to judge fairly whether the *argumentum ad misericordiam* is reasonable or not. In such cases it is best to suspend judgment. Of course, it is a good critical skill to be able to ask questions that express doubts that an appeal to pity should be taken to have more weight than it deserves. But one should not leap ahead and declare that where such critical questions can be raised in a text of discourse the argument in the text commits a fallacious *ad misericordiam* appeal.

The problem with this type of case, however, as already noted in case 4.7, is that the appeal to pity tends to be overwhelming. This is especially true where the overt, pictorial appeal to pity is in the form of a photograph or other visual presentation. The impact of the appeal to pity carries such weight that it alone can strongly influence the conclusion a not very critical respondent will form in the case. It is worth being careful with the appeal to pity, even if it is not demonstrably a fallacy.

The following is taken from a newspaper article reporting on the case of a man who was deported from Canada by the government because he had entered the country illegally. His wife, a Canadian citizen, had borne their second child just a month after the man had entered the country illegally. The reason the government gave for deportation was that the man had been convicted of car theft and breaking and entering while he had resided in the U.S.

Case 4.12

> A visibly grief-stricken mother, clutching her two-year-old son, bid a tearful farewell to her husband as he was deported from Canada yesterday. . . . [The mother sobbed,] "Why are they trying to destroy my family?"
>
> "Why are you crying mommy?" two-year-old Jimmy asked, just before the family shared one last embrace.
>
> "My boys are too young to understand what's happening," she said.

"What do I tell them tomorrow when they're asking for their father?" (*Winnipeg Free Press*, 27 April 1985, 1)

The article describes young Jimmy straining to stand on tiptoe to look out the window as his father is being driven away. It goes on to question why the government has refused this man a work permit and also deported him, pointing out that his wife will have to remain on welfare now that he has left the country. A large picture of the crying mother holding her baby accompanies the article.

According to the account given, the explicit conclusion is that the government's refusal to give this man a work permit and its deportation of him were questionable. However, the implicit conclusion is that the government should have taken the special circumstances of this man into account instead of deporting him.

The article is an appeal to pity, citing the particulars of the effects on his family of the man's forced departure. In order to evaluate the weight of the appeal to pity, more about the government's reason for the deportation needs to be known.

The article does tell us the government's reason and does provide some relevant information. Therefore, the appeal to pity is not unreasonable. However, since the article gives so much space to the appeal to pity, it could have given more space to the government's case or reasons for deporting this man. Perhaps the government was forced to deport him because of treaties or obligations with the U.S. government. At any rate, it would be relevant to know the government policy as it applies to this case and the reasons behind the policy.

The hasty or uncritical reader is likely to come away from this kind of case with a prejudged or one-sided view because the newspaper account has (a) not given all the relevant facts in enough detail to enable the reader to decide the issue fairly and (b) made such a strong appeal to pity that this alone may be enough to cause the reader to think that a terrible injustice has been done. Even so, it is better to resist the temptation to classify this type of case as an instance of the *ad misericordiam* fallacy. The best reaction is to pose critical questions that ask for more information about the particulars of the case.

Evaluating Appeals to Pity

In revising the textbook accounts of the appeal to pity, one needs to get away from the traditional approach of classifying the appeal into two categories—fallacious or non-fallacious. A categorization that provides for a gradation of cases would be better. For example, the following system of categorization is helpful.

1. *Reasonable.* Some appeals to pity, like cases 4.1, 4.2, 4.3, and 4.4 are emotional appeals in argument, but nevertheless they are reasonable and appropriate as moves of argumentation in their context of dialogue.
2. *Weak, but Not Irrelevant or Fallacious.* An appeal to pity in argument is open to challenge or critical questioning because it presents only one side of the issue. In case 4.6, for example, Bob's appeal to pity was relevant, but Helen needed to counter it strongly in order to deal with it effectively and critically.
3. *Irrelevant.* An appeal to pity may be irrelevant. For example, in a scientific inquiry an appeal to pity may be totally inappropriate to the extent of argumentation. In other cases the relevance of an appeal to pity may be marginal, and this may need to be pointed out by questioning its relevance in the dialogue.
4. *Not Enough Information Given.* Not enough information on the particulars of the case has been given, making it impossible to say definitively whether the appeal to pity is reasonable or not. The suspicion may be that the appeal to pity has been given undue weight of presumption in the case and that other important and relevant factors have been overlooked or suppressed. Often, the high impact of a pictorial appeal to pity is questionable, precisely because other relevant factors have been pushed into the background.
5. *Fallacious.* The appeal to pity is not just a weak argument open to critical questioning. It is used to prevent critical questions in subsequent dialogue by pressing ahead aggressively. In such cases, its use constitutes an unfair and deceptive tactic, and it is appropriate to make the charge that the offender has committed an *ad misericordiam* fallacy.

Using these five categories as a basis for evaluation involves a careful analysis of the given text of discourse and context of dialogue in a particular case. Careful attention should be given to the exact wording of the argument, discriminating between what has been said and what has been left unsaid in a given case.

This new approach redefines the problem posed by a questionable appeal to pity in argumentation. The problem is not so much one of deciding whether the appeal is a fallacious *ad misericordiam* as it is one of reacting to the appeal to pity in the right way. As a framework for a proper critical reaction to any appeal to pity in argument, the following checklist of seven factors should be kept in mind.

1. *Context.* What is the context of dialogue? Are emotions like pity relevant? Is it a critical discussion, or is the context an appeal for charitable action, for example? The appeal to pity must always be evaluated from a point of view of its contribution to, or obstruction of, the goals of dialogue. It is therefore necessary, as a first step, to identify the type of dialogue the participants in the argument are supposed to be engaging in.

2. *Weight.* How much weight should be given to the appeal to pity? Is appeal to pity central, as, for example, it would be in a charitable appeal to help starvation victims? Or is appeal to pity peripheral?

3. *Countering.* If the appeal to pity is central, then one must react to it strongly. This lesson was brought out graphically in case 4.7, where a film clip of a fox being electrocuted had an overwhelming impact. The case pictured was highly relevant to the main issue of the discussion and could not be brushed aside lightly.

4. *More Important Considerations.* If the appeal to pity is peripheral, show this by citing what are the main, or most important, factors in the case. For example, in the case of an appeal to pity used to support a plea to be excused from a rule, or in any other kind of excuse, what may need to be stressed are the main questions of whether the rule fits the given case, whether the excuse has been recognized by precedent, and so forth. Consideration of these more "factual" issues often takes the weight and emphasis off the emotional factor of pity.

5. *Balance.* If the appeal is an argument from consequences, balance out both sides of the equation. For example, in case 4.8, the case of the Non-Smokers' Health Act, the other side of the equation needed to be filled out by indicating the negative consequences of smoking. Cases like this are based on practical reasoning, where appeal to pity by citing possible long-term negative consequences for some group or sector of

the population is not out of place. But it can often be counterbalanced with opposite consequences.

6. *Improper Presumption.* The fallacious cases of *argumentum ad misericordiam* tend to be cases where the proponent has pressed ahead in a partisan attack, pushing forward a presumption aggressively in order to cover up the need to argue for it in a proper manner. Cases have been presented here in which presumptions alleging blame were packed in as presuppositions of loaded questions or a pictorial appeal to pity was used to push a presumption forward as taken for granted. Such cases are associated with begging the question and the fallacy of many questions.

7. *More Information.* In responding to many appeals to pity, it is necessary to request more information about the particulars of the case before any firm criticisms or reactions can be "pinned down." Instead of alleging "Fallacy!" right away, very often the best first reaction is to ask for more information.

The original, motivating question was: Should the appeal to pity be treated as a separate category of fallacy in its own right by the textbooks?

The answer is a provisional and guarded "yes." Plenty of evidence has been presented that the *ad misericordiam* is both a powerful and deceptively misleading tactic of argumentation well worth careful study and evaluation.

On the other hand, our treatment also suggests that it is misleading, in various ways, to think of the appeal to pity simply as a fallacious argument move. The problem is not that appeal to pity is inherently irrational or fallacious. The problem is that such an appeal can have such a powerful impact that it easily gets out of hand, carrying a weight of presumption far beyond what the context of dialogue merits and distracting a respondent from more relevant and important considerations.

While *ad misericordiam* arguments are fallacious in some cases, it is better to think of the *argumentum ad misericordiam* not as a fallacy (at least *per se*, or even most importantly) but as a kind of argument that automatically raises a warning signal: "Look out, you could get in trouble with this type of argument if you are not very careful!" It is a kind of argumentation that can be used in a fallacious way with such a powerful effect that one should have an attitude of critical vigilance in using it, or in responding to it. What is important is to react to it by asking questions that place the appeal in a context of dialogue where it can be properly evaluated.

5

ARGUMENTUM AD BACULUM

At first sight the *argumentum ad baculum* seems like such a transparently bad move in argumentation that no analysis of why it is a fallacy is needed. Appeal to threat of force or to fear as a move in a critical discussion, where both sides are critically examining the pros and cons of an issue in polite conversation, seems so radically out of place and underhanded that surely it should be categorically condemned as fallacious.

But if this were so, it would be hard to account for the "seeming validity" aspect of the *ad baculum*. Or, to put it another way, if the *ad baculum* argument were so transparently erroneous, how could it be such an effective and deceptive tactic in argumentation that it is worth studying as a logical fallacy? The answer to this question, advanced in this chapter, is that appealing to threats (where "threat" is defined broadly to include indirect threats and appeals to sanctions—for example, financial disincentives) is a nonfallacious move in some contexts of dialogue, especially in negotiation dialogue. Thus, whether an *ad baculum* is to be judged fallacious or nonfallacious depends on the context of dialogue. This makes the *ad baculum* a contextual error that requires a pragmatic analysis. The seeming validity can then be explained as arising from the dialectical shift from one context of dialogue to another during the argument. The *ad baculum* can seem plausible on the surface and can fool people as a fallacy because, in some contexts of dialogue, it is a nonfallacious kind of argumentation.

The Textbook Accounts

It would seem at a rough estimate that half of the logic textbooks that deal with informal logic treat the *ad baculum* specifically as a fallacy worth categorization and evaluation in its own right. Among the textbooks that do not treat *ad baculum* as a separate fallacy are Fearnside and Holther 1959, Fischer 1970, Michalos 1970, Johnson and Blair 1983, Govier 1985, Freeman 1988, and Waller 1988. Among those that do treat it as a separate category of fallacy, the *ad baculum* is characterized in three essentially different but overlapping ways: (1) appeal to force, (2) appeal to fear, and (3) appeal to a threat.

Copi (1986, 91) defines the *argumentum ad baculum* as "the fallacy committed when one appeals to force or the threat of force to cause acceptance of a conclusion." The problem with this account is that it is not clear that an appeal to force or the threat of force to cause acceptance of a conclusion is a fallacious kind of argumentation in negotiation. But Copi made the use of such appeals sound bad by characterizing them with phrases like "might makes right," "nonrational methods of intimidation," and "goon squads" (91).

The first of the two examples given by Copi (91–92) is arguably not a fallacy.

Case 5.1

> The lobbyist uses the *ad baculum* when he reminds a representative that he (the lobbyist) can influence so many thousands of voters in the representative's constituency, or so many potential contributors to campaign funds. Logically these considerations have nothing to

do with the merit of the legislation the lobbyist is attempting to influence. But they may be, unfortunately, very persuasive.

If the context of dialogue is that of a negotiation, if the lobbyist is arguing that he will deliver so many votes or dollars for commitment to support a piece of legislation, then reminding the representative how many voters or contributors he can influence is an appropriate and legitimate way to support his side of the negotiation. This "reminder" or "threat," however it is taken or interpreted by the representative or others, may be exactly the means for the lobbyist to fulfill his legitimate goal of the dialogue—namely to get the representative to agree to support the legislation in question. The lobbyist is saying, "Here's the deal," and if the offer is persuasive, that means it is a good and effective offer relative to the purpose it was designed to fulfill. Thus, it is somewhat disingenuous to call the lobbyist's reminder a fallacy because it has "nothing to do with the merit of the legislation" at issue. As Wreen (1987, 37) put it, in reference to this case, "[The lobbyist's] conclusion is not that the bill is a good one, but that the politician ought, from the point of view of self-interest, to support it." This could be a strong argument.

Indeed, according to the analysis of negotiation as a rule-governed interaction (Donohue 1981a, 111), threatening an opponent or an opponent's position in negotiation is a normal attacking move that obligates the opponent to respond in a certain way. Although Donohue states that threats are generally used as a last resort, he does not describe them as against the normative rules of good negotiation dialogue. In Donohue 1981b (279), promises and threats are said to be among the core concepts of bargaining theory and described as tactics that suggest "penalties or reward for performing or not performing an action." In Donohue's theory of negotiation as a type of dialogue, tactics are the techniques used by participants to carry out goals of bargaining in a negotiation situation, and threats are a species of attacking tactics. In this framework, there is nothing fallacious nor erroneous about threats *per se* in negotiation dialogue.

Whether a fallacy has been committed depends on what the representative and the lobbyist are trying to do. If they are discussing the merits of the legislation, then offers or threats to influence voters or funding are beside the point and make no contribution to the discussion. In that case, the lobbyist's "reminder" could be dismissed as a fallacious move on the grounds of its irrelevance to the issue of the discussion. But from the details of the example given by Copi, it is by no means clear, nor should it be taken for granted, that this is the case.

Kielkopf (1980) has rightly criticized the standard textbook account of the fallacy of *ad baculum* as inherently superficial. Citing Copi's analysis of the *ad baculum* as "the fallacy committed when one appeals to force or the threat of force to cause acceptance of a conclusion," Kielkopf (1980, 2) poses the following case: "To get people to accept the conclusion that they should not drive under the influence of alcohol it is perfectly reasonable to threaten them with loss of driving privileges." The problem, Kielkopf notes, is that a student can use a definition like the one given by Copi to condemn arguments essentially like that in the loss of driving privileges case above as instances of the *ad baculum* fallacy. Kielkopf rightly criticizes this approach as not only superficial but based on a kind of dishonesty, because it involves condemnation of a whole class of arguments some of which are not only nonfallacious but are both common and reasonable ways of persuading people to act one way by pointing out the negative consequences of their acting differently. By encouraging students to classify such arguments *holus-bolus* as fallacious *ad baculum* arguments, the standard textbook treatment could be said itself to commit or rest on a kind of fallacy. At any rate, important distinctions have been overlooked.

Michalos (1970) treats "force" and "fear" as two separate categories of fallacies. Under the fallacy of irrelevant appeal to *force*, Michalos (51) includes examples of arguments like the following: "Either I'm right or you don't get the car tonight; therefore I'm right." Under the fallacy of appeal to *fear*, Michalos (58) gives the following example (paraphrased below):

Case 5.2

An insurance salesman, who has come by to "inform" a client of the advantages of buying a policy proceeds to describe the plight of another man who did not have a policy. This man's house burned down, and he then suffered terrible hardships, according to the salesman, who goes on to describe these hardships in graphic and lengthy detail.

Michalos (58) comments that the salesman uses fear to drive his respondent to the conclusion that he or she needs a policy, without, however, appealing to force at any point. Michalos calls the appeal to fear the *argumentum ad metum*, in contrast to the appeal to force, which he calls the *argumentum ad baculum*.

The problem with this approach to classification is that force does not seem to be essential to the *argumentum ad baculum*, as case 5.1 shows. On the other hand, the emotion of fear does seem to be closely, even if not essentially, part of the *ad baculum*. An argument that appeals to fear by making a threat but makes no appeal to force—for example, where bankruptcy or some other unpleasant consequence is threatened—is a fairly central type of *ad baculum* argument.

Whereas Copi calls the *ad baculum* "appeal to force," Engel (1976, 130) calls it "the fallacy of appeal to fear," defining it as "an argument that uses the threat of harm to advance one's conclusion." But he adds that an "all-out threat," like pointing a gun and saying, "Your money or your life," is not a fallacy because it is not an argument. The claim that this type of threat is not an argument is, however, flatly denied by Woods (1987). Woods considers the case he calls "the heist," where one individual points a pistol at another, saying, "Your money or your life." According to Woods, there is no fallacy in this case—"it is a good prudential argument."

Woods makes a convincing case that the heist performance is an argument, based on what he calls *a prudential argument*, where the proponent proposes an action that he claims, "cohere[s] with the interest of" the respondent to whom the argument is directed. According to Woods (1987, 344), there can be both good and bad prudential arguments.

Woods' skeptical questioning of the standard treatment of *ad baculum* in the textbooks challenges a long-standing tradition by raising doubts whether the *argumentum ad baculum* is really a fallacy at all. In a series of four articles, Wreen (1987, 1988a, 1988b, and 1989) carried this criticism of the standard treatment even further. According to Wreen (1987, 31), the textbooks "punk out" by using a few stock examples that are not subjected to a serious enough analysis as arguments that occur in a context. After analyzing many of these examples himself, Wreen concludes (1988a, 438) that most *ad baculums* turn out to be "fairly strong," once they are "reconstructed and carefully evaluated." Wreen is even led to doubt whether the *ad baculum* fallacy exists, or if it does, he doubts whether any of the textbook treatments have ever defined it in an adequate way.

Attempts to give an account of the *ad baculum* fallacy after the pattern of Copi or Engel fail to recognize and cope with the fact that this argumentation can be nonfallacious in contexts like negotiations or prudential arguments. A more sophisticated account is given by Damer

(1980, 91), who defines "appeal to force or threat" as the fallacy of
attempting to persuade by threatening undesirable consequences instead of
giving evidence for one's view. Damer recognizes, however, that pointing
out bad consequences of a proposed course of action need not be fallacious.

> There is nothing wrong, of course, with pointing out the conse-
> quences of a particular course of action. In fact, if certain
> consequences are a natural outcome of an action, calling attention
> to them might be very much appreciated. In some such cases, being
> aware of the consequences of an action might even cause one to
> alter one's course. However, if an arguer tries to force another to
> accept the truth of a claim or the rightness of an action by
> threatening some undesirable action, then the arguer is guilty of
> using an irrelevant appeal.

The recognition of this subtlety is a favorable aspect of Damer's way of
presenting the *ad baculum*, but he could have gone further. Not only is
pointing out consequences of a proposed course of action nonfallacious
argumentation; even threatening someone with bad consequences should
he or she do or not do something is a nonfallacious kind of argumentation
in some contexts. But Damer doesn't go this far and doesn't offer an analysis
of how to sort the fallacious from the nonfallacious cases.

   Clearly, argumentation from consequences of a proposed course of action
is woven in closely with the *argumentum ad baculum* and is an essential
ingredient for understanding its structure as a logical fallacy. But there
seems to be some confusion and uncertainty in the logic textbooks
concerning the nature of this connection. Cederblom and Paulsen (1982,
102) describe the fallacy of "appeal to force" as a particular instance of the
more general fallacy of "appeal to consequences." Contrary to Damer, as
quoted above, who sees pointing out the consequences of an action as a
legitimate kind of argumentation, Cederblom and Paulsen portray appeal
to consequences as a fallacious species of argumentation. By their account,
this fallacy occurs where someone accepts an appeal to force, or to some
other emotion like pity, "not because of evidence but in order to bring
about certain consequences—avoiding harm to yourself or others." They
give the following example.

Case 5.3

> If you opposed gun control you'd have a lot better chance of being
> elected. Why don't you reconsider your position on that issue?

The first problem with this case is that it seems more like an *ad populum* appeal then an *ad baculum* argument. But there is no reason why both types of appeal cannot be combined in one case. And in fact this case is interesting in its own right because it would appear to be a borderline case where the two tactics have been combined in one argument.

But the second problem is that it is not clear that case 5.3 is a fallacious argument. Note that the speaker's recommendation is put in the form of a question. It is not even an assertion in the form of a proposition inferred or put forward as a conclusion. Moreover, it seems like it could be a reasonable question, for all that is known from the given context of dialogue and text of discourse. After all, a politician who goes against popular opinion probably should reconsider her position on the issue. Note that the speaker is not forcing her to change her position, or even insisting or asserting that she must or should change it. The speaker is only asking, "Why don't you reconsider your position on that issue?" Of course, another interpretation is that there is a subtle threat of harm or bad consequences in the statement that "you'd have a lot better chance of being elected," the implication being that if the politician doesn't change her position, she will likely not get elected. But is such an argument from consequences inherently fallacious? It is far from obvious that this is the case. And it seems the more reasonable presumption, as suggested by Damer above, that argumentation from consequences, even bad or harmful consequences, is not in itself erroneous or fallacious.

In summary, then, while the textbook accounts raise some interesting questions about the *ad baculum* as a type of argumentation, there is little agreement on how they describe the fallacy, or even on how they label and classify it. There would appear to be even less consistency on how the fallacy is to be analyzed as an error or fault of good argumentation. And there do seem to be fundamental confusions about whether cases coming under the heading of *ad baculum* are really fallacious or not. Interesting puzzles are posed, but the textbook accounts are lacking any underlying theory that could explain or justify their intuitions that the *ad baculum* is a serious logical error worth studying and guarding against.

Defining *Argumentum Ad Baculum*

As noted above, the textbooks vary on how to characterize the *argumentum ad baculum*. This means that prior to the job of analyzing or judging whether and how this kind of argument is a fallacy, it must be decided what it is. Three possible ways of defining it were noted above—appeal to force, appeal to fear, and appeal to a threat. However, there are disjunctive possibilities for defining it as well—Copi defined it as appeal to force or threat of force, and Damer defined it as appeal to force or threat. It could also be defined inclusively as appeal to force or a threat or fear. Many combinations are possible for different definitions of the *ad baculum*.

These alternatives make a real difference. For example, it is possible to have an appeal to force made in a diplomatic negotiation without either side being fearful, without even the intent to evoke fear. And it is possible, in some cases, to appeal to fear without making a threat, or without threatening the use of force (see case 5.2 above). For example, one person might try to influence another to take a certain route by saying that snakes are known to frequent an alternative route, thus exploiting the second person's fear of snakes to influence his or her choice. This would certainly be an appeal to fear but would not necessarily involve making a threat.

As a starting point, it would seem plausible that force and fear are more incidental or optional elements of the *argumentum ad baculum* and that the essential element is that of a threat. But there remains room for doubt.

How should the *argumentum ad baculum* be defined? Should it require a threat of force? Or only a threat? Or should an appeal to fear count as well? Or would even an allusion or reference to unpleasant consequences in an argument be enough?

Wreen (1988a) thinks that there can be an *argumentum ad baculum* that "involves neither force nor the threat of force," and he presents an interesting example designed to prove the point (432).

Case 5.4

> If you and I are taking a stroll, and you inadvertently wander onto a patch of quicksand, I might stand by as you scream for help, musing aloud about the vicissitudes of fortune and how much I've always wanted that diamond ring you're wearing, a diamond ring bequeathed you and not me, I remind you, merely because of your good fortune. I think you'd understand me. I'm arguing:
>
> > If you don't give me your diamond ring, I won't help you get out and you'll die.
> > Your dying is a great evil you would suffer.

Therefore, you ought (from the point of view of self-interest) to throw me your diamond ring.

According to Wreen (432), neither force nor violence is involved in this case, and neither is it clear that I am threatening you by taking unfair advantage of your precarious situation.

But does case 5.4 constitute an instance of the *ad baculum* fallacy? The type of dialogue in case 5.4 is that of a negotiation. It is certainly an unfair tactic of negotiation that is used, and it is not very nice for many reasons. It's rude, selfish, uncaring, unpleasant, and possibly even illegal, as an action or omission to act. But is it a fallacy? The answer here is unclear. In the example, I am trying to persuade you to make a deal, to offer the ring, but I am not trying to persuade you to accept some proposition as true or false, to take on a commitment representing a point of view. The failure is not one of the abuse of some rule or procedure of a critical discussion the two of us are supposed to be engaged in. But perhaps it could be an *ad baculum* fallacy all the same if it is a violation of some rule or proper procedure of negotiation dialogue.

But even if case 5.4 is not an *ad baculum* fallacy, is it an *ad baculum* argument? No clear or obvious answer is forthcoming for this interesting borderline case. As a strictly prudential argument in a negotiation, given that no threat is involved, case 5.4 does not really seem like an *ad baculum* argument. Of course, superficially, it looks like an *ad baculum*, but as Kielkopf (1980) and Woods (1987) have rightly counseled, a harder look at superficial appearances needs to be taken in such cases. Until the elements of a genuine *ad baculum* argument have been identified it would seem premature to rule on case 5.4.

Another puzzling thing about case 5.4 is the question of whether a threat is involved or not. Wreen is cautious here, writing, "I'm not sure I'm *threatening* you as I babble on while you are sinking in the quicksand" (432). But this interpretation, depending on the context, could be quite naive. I am not directly or overtly expressing a threat perhaps, but threats are often indirect speech acts. On a deeper interpretation of my speech act, what I say is threatening to you, and I know that. Hence, indirectly I am, in effect, threatening you by saying that you will get no help (and consequently die) if you do not throw me the ring.

This case not only challenges the reader to state explicit requirements for what constitutes an *ad baculum* argument; it also challenges the reader to define the concept of a threat. It does not prove conclusively, however, that it is possible to have a genuine *ad baculum* argument that involves no threat.

A threat is more than just a statement to the effect that unless one does (or fails to do) something, some bad or painful consequence will occur. It also expresses a commitment on the part of the proponent of the threat to do (or fail to do) whatever is required to bring about or expedite this painful consequence. A threat involves the proponent saying or implying that she will take on a certain commitment to bring about or contribute to this outcome and that she is or will be in a position to do this effectively. A threat requires (a) some words, gestures, messages, etc., conveyed by the proponent to the respondent, (b) a situation or context in which this message may be interpreted, by the respondent and others as well, as expressing a threat, and (c) a fitting together of these two elements into a coherent picture that enables third-party observers to infer or judge the nature and seriousness of the speech act as a possible threat made to the respondent by the proponent. Judging or interpreting a speech act as a threat is clearly a pragmatic task that involves essential reference to the specific circumstances of a given situation.

Does the speech act in case 5.4 amount to a threat or an offer? If interpreted as an utterance that says, "I am not going to help you escape in this threatening situation unless you give me the ring," it amounts to a threat. If interpreted as an utterance that says, "I am going to offer you a deal—I'll come to your assistance if you'll give me the ring," it amounts to an offer. But it could be both. It is an offer backed up by what amounts to a threat or an appeal to the danger of a threatening situation.

Case 5.4 depends on presumptions about the context of the situation. It appears to presume, for example, that there is absolutely no danger in my rescuing you from the quicksand. Supposing that there could be some risk to my life in attempting to rescue you from the quicksand makes the proponent's speech act in case 5.4 appear more like an offer and less like a threat. It still seems like a selfish and reprehensible offer but less a threat, and perhaps also less like an *ad baculum* argument.

Should Appeals to Fear Be Included?

Should appeals to fear be counted as genuine *ad baculum* arguments, even if no threat is involved in the argument? One thinks of the common case, for example, where an orator plays on the fears of his audience by suggesting or alluding to some ominous possibilities to which this particular audience is sensitive. Of course, one can also appeal to fears that could be described as rational, rather than emotional or "hysterical," so appeal to fear in argumentation is not necessarily fallacious. But when an argument does appeal to fear but makes no threat, should it be classified as an *ad baculum* argument? Or, like Michalos (1970, 51), should it be put in a separate category?

The difference between the two types of argument at issue is that in the threatening argument the speaker is expressing a readiness to take steps that will bring about the bad outcome feared by the hearer, whereas in the argument by appeal to fear (where no threat is involved) the speaker is not stating or implying any intention to take part in bringing about this bad or fearful outcome.

Of course, any appeal to the fear in an audience will be inherently threatening to that audience. But arguing in a way that is threatening to your hearer and actually threatening your hearer—that is, making a specific threat against him—are two different things. The question is whether an argument that appeals to fear but does not make a threat to the respondent can correctly be classified as an instance of the *argumentum ad baculum*. Arguments that appeal to fear often appear to be very suspicious but, when examined closely, may turn out to suffer from faults other than the *ad baculum* fallacy.

There is no question that fear is an emotion systematically exploited by advertisers in designing commercials. Fear is a basic human motivation, and it would be surprising if advertisers, like filmmakers, novelists, and playwrights, did not appeal to it. Two graphic cases are described by Clark (1988, 111), labeled (a) and (b) in 5.5.

Case 5.5

(a) In what might be considered the ultimate in shock advertising, the father in a Pakistani commercial finishes his dinner, lights a cigarette, suddenly clutches his chest and falls to the floor. In hospital doctors struggle to save his life, but fail. The doctor pulls the sheet over the man's face, turns to the camera and tells the viewers, "This could be you if you don't give up smoking." (b) A UK print advertisement for a private health-care company showed a caricature of the Grim Reaper standing over a row of occupied hospital beds with the headline "When you're old, the last place you want to be is in hospital."

According to Clark (111), the commercial in (b) was later censured by the Advertising Standards Authority because it violated a British Code of Advertising Practice ruling that banned "unjustifiable use of fear tactics."

Because almost everyone would agree the aim of persuading people to give up smoking is a good one, case 5.5a would not be used in logic textbooks as an example of the *ad baculum* argument as an obviously fallacious tactic. To pin down what might be considered fallacious about case 5.5b as an *ad baculum* appeal is also not so easy, because in fact dying in hospital of smoking-related diseases like emphysema (chronic obstructive lung disease) can be extremely unpleasant, and it is very worthwhile to warn people about it. And if scaring people is an effective way of getting them to give up smoking, then even a considerably frightening appeal to the harmful consequences of smoking, portrayed in a graphic and pictorial manner, could be justified. Or, at any rate, it would not do to presume without further argument that such an appeal to fear can be automatically ruled out or labelled as a fallacious *ad baculum*.

The commercial in case 5.5b was banned on the grounds that it violated a rule of advertising practice that outlawed "unjustifiable use of fear tactics." But the word "unjustifiable" suggests that not all appeals to fear in commercial advertisements are considered illegitimate by the standards of advertising practice. And second, one needs to ask on what grounds, or for what reason an appeal to fear is judged "unjustifiable," for a case of bad advertising practice is not necessarily a fallacy.

This case raises the question, What is the purpose of a commercial advertisement? Presumably it is to persuade buyers to action or, in particular, to buy a particular brand or product. But are commercials supposed to persuade potential buyers rationally, by giving reasons why a course of action is recommended as prudent or why their product is good, useful, a good buy, etc.? Or are commercials not to be held to such lofty standards of rationality—can they, often legitimately, just be ways of drawing attention to a brand or product, or of creating a favorable emotional ambience around that product? It is not easy to say exactly what the purpose of a commercial should be, but it seems more plausible that there should be some latitude in allowable purposes, and that in fact many commercial advertisements and promotions create a climate or scene that evokes emotions rather than engaging in explicit rational argumentation. Applying these considerations to case 5.5a, it is not obvious that it is an instance of the *ad baculum* fallacy. Of course, it could be in bad taste, or frightening enough to offend many viewers—think of children, in particular. But neither of these shortcomings need imply that a logical fallacy has been committed.

What can be said is that this advertisement would have been more "rational" if it had cited statistics on emphysema and other smoking-related diseases, or gave other medical reasons for the prudential course of action, stopping smoking, as a conclusion. Such a criticism by no means demonstrates that the failure of the argument in case 5.5a to do such things, instead of relying on a pictorial appeal to fear, makes it fallacious. For what is unclear, or has not been established, is that the purpose of a commercial advertisement is to persuade by giving "objective," "factual" reasons, as opposed to creating a scene, ambience, or drama that will generate an emotional response in viewers that will in turn cause actions.

The trickiest appeals to fear in argumentation are the indirect appeals to subliminal fears, those unstated and unarticulated fears that lie under the surface of a respondent's commitments. This kind of appeal can be very effective if it is directed to a powerful fear that a person or group of individuals may have without even recognizing explicitly that they have it.

During the 1988 federal election in the United States, Republican campaign advertisements attacked the liberalism of Michael Dukakis by showing convicts walking through a "revolving door" in and out of the penitentiary in Massachusetts. In Maryland, a Republican fund-raising letter included photographs of Dukakis and Willie Horton, a furloughed black rapist, accompanied by the headline "Is This Your Pro-Life Family Team for 1988?"[1] Horton, while on furlough from a Massachusetts prison, where he had been serving a life sentence for murder, tortured a Maryland man and raped his fiancée. The suggestion of these campaign messages was that black-on-white crime would increase if Dukakis became president, because Dukakis's liberal attitudes made him "soft on crime."

This tactic was an appeal to fear of the *ad baculum* type, the argument being that if Dukakis were to be elected president, dangerous criminals would be let loose, a menace that could affect the average citizen with horrifying results. By appealing to fears of the average citizen, this tactic proved very effective.

1. CP Report, "GOP Campaign Plays on Racial Tension, Crime," *Winnipeg Free Press*, 31 October 1988, 3.

These advertisements allegedly appealed also to racial tensions by aiming at a "sensitive nerve" for white middle-class Americans.[2] Although Republicans denied having racist-directed intentions in making up the advertisements, it was inferred by critics that these messages were so effective because they appealed to white fears relating to the growth of violent and powerful black gangs in big cities like Los Angeles and New York.

Case 5.6

> Michael Kinsey of the *New Republic* magazine wrote in a recent column that the Republican campaign "taps into a thick vein of racial paranoia that is a quarter inch below the surface of the white American consciousness."
>
> Ronald Walters, a political scientist from Washington's predominantly black Howard University, says he believes race is the "central fact" of U.S. society and politics.
>
> "Many American whites still see every black man as a potential mugger, rapist, drug dealer, or murderer," he said in a recent interview. "These ads tap into those fears and convey the message that if Dukakis becomes president, blacks are going on a rampage of raping and killing."[3]

The particular audience for which this appeal was thought so effective were the millions of northern blue-collar white voters who had deserted the Democratic party to vote for Ronald Reagan in the two previous federal elections.

On the surface, there is nothing wrong with the Republican criticisms of Dukakis's policy on crime, and at least in principle, it was not inappropriate for them to cite a particular case where as Governor of Massachusetts he had allegedly been responsible for the parole of someone who subsequently committed a crime. The implication is that Dukakis's policy was at fault, and in the context of an election campaign, it is legitimate to cite such a case where one's opponent was allegedly at fault. Mr. Dukakis then had the opportunity to reply to this allegation.

But beyond this the Republican advertisements appealed to fear as an

2. Ibid.
3. Ibid.

election campaign tactic. Evidently, this tactic worked, but the reasons why it was so effective are to be sought in the message that the advertisements implied or suggested, beneath the surface of their ostensible criticism of an opponent's viewpoint. Under the surface, the tactic appealed to fears about crime and also perhaps to other powerful but unarticulated fears, presumably felt by a significant population of voters (at least, according to critics).

If the critics were right, the tactic used in these advertisements was tricky, subtle, indirect, and powerful. But precisely what is wrong about it as a logical fallacy is not easy to pin down. It does not seem to be a failure of relevance, for the Dukakis policies on crime were a relevant and legitimate issue in the campaign. What then, exactly, was the fallacy? If it was not the *ad baculum* fallacy, was the argument even an *argumentum ad baculum*? Perhaps the job of locating and analyzing the faults and tactics of this type of argument may be hampered more than facilitated by classifying it as an *ad baculum*.

There are two sides to this question of whether the appeal to fear in an argument that does not contain a threat should be classified as an *ad baculum*. On the one side it should be said that the appeal to fear is a key aspect of *argumentum ad baculum*, even where a threat is involved, because it is the appeal to fear that is the basis of the tactic of masking a lack of evidence. But on the other side it should be said that indirect appeals to hidden subliminal fears appear in all kinds of difficult, subtle cases where clever advertisers and propagandists have decided that logic may not be much help in sorting out what is supposed to be wrong.

It is easy to be pulled both ways here. It would be much simpler and clearer to stick to the central type of *ad baculum* where a threat has been used and concentrate on the analysis of this tactic as the *ad baculum* argument. But, on the other hand, many appeals to fear do appear to work as effective tactics of argumentation because they are threatening to the respondent, even if no specific threat has been made by the proponent. And the textbooks do include such cases as *ad baculum* arguments, even though citations of this type would appear to be a minority of cases.

A compromise would be to distinguish between two types of *argumenta ad baculum*. The central type involves a threat made by the proponent, and the peripheral or derived type involves an appeal to fear without requiring the threat.

When Is Using a Threat Fallacious?

One thing to watch out for in approaching the *ad baculum* is the temptation to infer that because an argument move of this sort is a threatening or even brutal or repulsive use of force, it is a fallacy. Appeals to force can be impolite, immoral, illegal, ungracious, or wrong for all kinds of reasons without necessarily being fallacious. Committing a fallacy involves the use of a technique of argumentation to go against legitimate goals of reasoned dialogue—it means that there is an underlying, systematic error or failure in the reasoning used to carry out goals of dialogue appropriate for the given case. Too often the texts simply presume that given a brutal or repulsive or morally repugnant appeal to force, the students or readers of the text will need no further persuasion to classify it as a fallacy, as something that cannot ever be rationally justified or defended, and that therefore can be quickly condemned.

Consider the following case (invented), where two representatives of hostile countries are engaged in diplomatic negotiations, and one of them appeals to the threat of force.

Case 5.7

> As the Syrian Defense Minister, I would like to make it clear to you, the Israeli Defense Minister, that there will be no easy war or victory if Syria is attacked by Israel. The U.S.S.R. has restored all the equipment Syria lost in Lebanon, and much more besides. We have fifty new MIG-23s and MIG-25s. We have new Soviet antiaircraft missiles, ground-to-ground missiles, and a sophisticated, electronic command-control system. We now have 600 aircraft, 3,600 tanks, and more than half a million men in our armed forces. We also have a defense agreement with Lebanon, and can count on their support if we are invaded.

On seeing this direct and somewhat brutal appeal to force, it could be easy to rush to the conclusion of classifying it as a fallacious *ad baculum*. But the context is that of diplomatic negotiations between two hostile countries. The Syrian defense minister is informing his respondent of the Syrian position and at the same time warning him about the Syrian capabilities for a military response. In the situation, of course, this warning and giving of information can quite rightly be interpreted as a threat. But diplomatic negotiations at this level and of this type are normally (and not necessarily illegitimately) based on sanctions and threats of reprisals or countersanctions of one sort or another. Such Machiavellian moves of bargaining and threatening are perhaps regrettable and unpleasant, but it is a big leap to automatically classify them as fallacious *ad baculum* arguments.

What is the conclusion of the argument in case 5.7? The conclusion is that there will be no easy war or victory if Syria is attacked by Israel. The argument may be interpreted as a threat or a warning, and it is certainly an appeal to force. But it is not a fallacious *ad baculum* since, as far as is known from the given context of dialogue, the appeal to force is not used to distract the other arguer from more relevant considerations. Nor is there other clear evidence from the given text of discourse that the threat is being used as an illicit tactic of argumentation or error of reasoning to block the goals of the negotiation dialogue.

To be considered a fallacy, not only must a threat be made; the threat must be used (illicitly) to try to persuade or dissuade someone in argument. The use of threat may constitute an *ad baculum* argument, but in order for the *ad baculum* argument to be an *ad baculum* fallacy, the threat must be used in some type of reasoned dialogue, like a persuasion dialogue, in a way that is not legitimate or appropriate in conducting that dialogue.

The following case was a newspaper report covering the essentials of a criminal trial where the defendant was charged with obstructing justice. The charge against him was that he had used a threat to try to keep another man from testifying against him.

Case 5.8

Broken-Legs Threat Didn't Warn Witness to Shut Mouth, Judge Rules

A man who wanted someone to break the legs of a Crown witness wasn't necessarily trying to obstruct justice, a judge ruled yesterday.

Mr. Justice Aubrey Hirschfield acquitted Harry Hartwick, 37, after ruling his threat wasn't aimed at keeping another man from testifying against him.

Hartwick was charged with willfully attempting to obstruct justice by using threats to dissuade Dave Heaney from giving evidence.

Ron Hartwick said his brother offered him $500 in August 1984 to break Heaney's legs and repeated the offer several times. Heaney was a witness in a trial last year in which Harry Hartwick was acquitted of assault, court was told. He said that incident, in February 1984, involved himself and his sister who had previously lived common-law with the accused.

Crown attorney Jannine LeMere said the only logical conclusion that could be drawn from Hartwick's offer to his brother was that he wanted to keep Heaney from testifying.

However, the Court of Queen's Bench judge agreed with defence lawyer Catherine Dunn who argued there was no evidence the threat was aimed at preventing Heaney from testifying.

Hirschfield noted her client could have had other reasons to be angry with Heaney.

Heaney said he didn't feel very good when he learned of the threats against him.

"I think he's capable of doing it," Heaney told yesterday's hearing.

However, Art Wooden, a person who heard Hartwick make the initial $500 offer to his brother, said he thought the man was joking.

"I don't think Harry was really serious about it," Wooden said. "I think he was more or less joking around."

Ron Hartwick said his brother told him he couldn't break Heaney's legs himself because a court ordered him not to communicate with Heaney.

He denied making up the story to get revenge because his brother had an affair with his former wife four years ago. He said if he wanted revenge, he could have gone to the police with his story.

Instead, he said he contacted Heaney who called police about the threat. (*Winnipeg Free Press*, Local, 5 April 1986, 3)

There are two questions at issue in this case that need to be distinguished. One question is whether Harry Hartwick had threatened to break Dave Heaney's legs. The other question is whether he used this threat to dissuade Dave Heaney from testifying against him.

Could it be said that Harry Hartwick was guilty of committing an *ad baculum* fallacy (turning now from legal to logical questions)? The answer is that if Harry Hartwick was guilty of using a threat to obstruct justice by preventing Dave Heaney from testifying, then he committed an *ad baculum* fallacy. The reason is that according to such an account of what Harry Hartwick was doing, he was using the threat of force, or appeal to force, to block Dave Heaney from taking part in a legitimate dialogue—in this case court testimony—and the threat of force was an illegitimate way of trying to influence Dave Heaney. In this case it would have been an attempt to block the legitimate progress of a dialogue by an improper means.

This case is important because it illustrates a necessary component of a genuine *ad baculum* fallacy. The appeal to force or the threat of force must not only exist; it must also be an appeal that is used to persuade a respondent to do something or to accept a conclusion in a manner that is inappropriate for the context of dialogue that is supposed to be taking place. To be a fallacy of *ad baculum*, the threat must not only be made to a respondent and must not only be wrong or illegitimate; it must also violate a rule of dialogue, or go against goals of the type of dialogue that the speaker and respondent are supposed to be engaged in.

Judging whether an *ad baculum* argument is fallacious or not is a job that is highly sensitive to the context of dialogue the participants are supposed to be engaged in. In a critical discussion, each participant has an obligation to prove to the other participant that his or her point of view is right. The participant persuades the other participant by presenting convincing arguments based on premises that the other participant will accept or is already committed to. The goal is to resolve a conflict of opinions. In this context, threatening another participant serves no useful purpose. Such a move would normally be out of place and highly suspect because it is not coherent with the goals and usual procedures of this dialogue type.

One common use of a threat in a critical discussion is to distract an opponent either by arousing emotional responses or by applying pressure instead of properly fulfilling a burden of proof. Such tactics can rightly be identified with the *ad baculum* fallacy. And indeed, van Eemeren and Grootendorst (1987, 285) describe the *argumentum ad baculum* as a violation of the rule of critical discussion, "Parties must not prevent each other from advancing, or casting doubt on standpoints," insofar as it puts pressure on an opponent. Any threat made in order to block the goals or proper procedures of a critical discussion would clearly be illicit and highly suspect.

But the situation is quite different if the threat occurs in a negotiation dialogue. Threats are common, even normal, in negotiation dialogue and can play a legitimate part in contributing to its goals. Explicit threats can be obstructive in negotiations and may in many cases actually not be allowed at all. But part of the cut and thrust of hard bargaining is to level sanctions against proposals that are not perceived as favorable to one's side—and such warnings about the implementation of sanctions are in fact subtle threats, conveyed by indirect speech acts. This aspect of negotiating, while it has boundaries that can easily be overstepped by excessive or inappropriate threats, is generally accepted as a tolerable and even legitimate form of bargaining.

The appropriateness of threats in negotiation dialogue becomes easier to accept as a thesis once it is understood that a threat does not necessarily entail an appeal to force or violence. A threat can also be an indirect speech act where the speaker suggests that unless the hearer does something, or refrains from doing something, the speaker will do something that is not in the hearer's interest. For as well as harsh threats, it is possible to have soft threats, where the action threatened is not violent, illegal, or any infringement of the freedom of the hearer. Indirect threats often exert subtle pressure without using coercion. Provided one defines the concept of a threat in a broad enough way, so that it includes soft threats as well as overt, direct threats of a harsh or violent sort, not every threat is required to have an element of menace or violence.

Such a wide concept of threat is presumed by the legal definition given in Saunders 1970 (190): "A threat is only an intimation by one to another that unless the latter does or does not do something the former will do something which the latter will not like." This legal definition captures the idea of an indirect threat. But of course it is also possible to have overt or direct threats that are expressed in stronger language than that of an "intimation."

Direct *Ad Baculum* Arguments

The direct *ad baculum* argument is characterized by the speaker's making a direct threat to the hearer. But what is "making a threat"? The answer proposed herewith is that making a threat is first a type of speech act. A threat is not a particular locution, not an imperative or a "prescriptive" rather than "descriptive" sentence. A threat can be expressed in the form of a question, an assertion, or an imperative.

What is essential to the making of a threat is that a speaker convey an undertaking to carry out a particular intention and that a hearer recognize this declaration of intention and understand its communicative force as a commitment of the speaker. Further, in order for a speaker to make a threat to a hearer, there must be some event that the hearer takes to be bad, or not in his or her interests; the speaker conveys the idea to the hearer that the speaker will see to it that this event occurs unless the hearer carries out some action. According to Wreen (1988b, 93), proper reconstruction of an *ad baculum* argument involves two linked premises, each of which is concerned with evaluation of an action, and a conclusion that is an 'ought' statement.

The speech acts of warning and threatening are basically different. Conditions necessary and sufficient for the act of making a threat can be given as follows.

1. *Preparatory Conditions:* The hearer has reasons to believe that the speaker can bring about the event in question; without the intervention of the speaker, it is presumed by both the speaker and the hearer that the event will not occur.
2. *Sincerity Condition:* Both the speaker and the hearer presume that the occurrence of the event will not be in the hearer's interests, that the hearer would want to avoid its occurrence if possible, and that the hearer would take steps to do so if necessary.
3. *Essential Condition:* The speaker is making a commitment to see to it that the event will occur unless the hearer carries out the particular action designated by the speaker.

According to this analysis, threatening is not a stronger kind of warning. It is not a warning of any kind. Making a threat is quite different from giving a warning, because making a threat is an attempt by the speaker to get the hearer to carry out an action. Moreover, it is based on the speaker's willingness to intervene and actually bring about the event in question.

Interpreting whether an utterance is really a threat or not is least difficult in the case of a direct threat, all the elements of which are present in a sequence of four frames in the cartoon *Blondie* (King Features Syndicate, Inc., 1973). A formidably large salesman takes out a bottle of window cleaner from his suitcase as Dagwood answers the door.

Case 5.9

> Salesman: I'm selling this window cleaner. And I'm not a guy who likes to fool around. Either you buy it, or I'll punch your lights out!
>
> Dagwood (walking back into his living room after buying two bottles of window cleaner): He has a very persuasive sales approach.

This case has the following elements of the *ad baculum* fallacy. First, the salesman makes a direct threat. Second, the salesman is supposed to be engaged in a persuasion dialogue where he uses arguments to convince Dagwood that this window cleaner is something he should buy because it is a good or useful product (or for whatever persuasive reasons the salesman can present). And third, in this context of persuasion dialogue, the threat does not constitute a good (relevant) reason for buying the window cleaner.

There is also a fourth element present, at least arguably. The threat may be a good prudential reason or argument basis for Dagwood's buying the window cleaner (to prevent himself from harmful consequences of failure to buy it). And this may be a good reason or argument in some context of dialogue, like a negotiation dialogue, where both parties are threatening each other with sanctions. But it is not a good reason in the type of dialogue which the salesman is supposed to be engaged in with Dagwood—a persuasion dialogue. In this latter type of dialogue, it is an inappropriate move that interferes with or even prevents the dialogue from carrying on in a normal or reasonable manner toward its goals. The "sales pitch" appropriate for this type of exchange is cancelled altogether or made unnecessary.

Hence, there are three, or arguably four, minimal elements of a simple *ad baculum* fallacy present in this case. Note, however, that the element of fear would appear to be optional or unnecessary in this case. Dagwood does not necessarily have to exhibit fear for the case to constitute an instance of the *ad baculum* fallacy.

Of course, one could argue that fear is implicit. Dagwood does not have to exhibit a hysterical fear, or an actual emotion of fear, for the argument to be an *ad baculum* appeal, for the salesman's tactic is based on an appeal to rational fear. Prudentially, according to the salesman's argument, the rational conclusion is to buy the window cleaner; the consequences of buying are less harmful than the consequences of not buying. So one could argue that a rational fear, or motive to avoid negative consequences, is a part of the salesman's argument.

At any rate, this case is theoretically interesting for the study of *ad baculum* because it contains the three minimal elements required for a simple, direct *ad baculum* fallacy. Note also that while it definitely contains a direct threat, it is less clear that it contains or requires an element of fear on the part of the respondent or that the appeal is directed explicitly to the emotion of fear. This type of case may be an emotional fallacy in some derived sense, but it does not require that the participants actually exhibit heightened emotions or behave in an especially emotional way.

This case then could be described as a classic and simple, streamlined instance of the *argumentum ad baculum*. There is nothing subtle about it. It is not meant to be a seriously deceptive case or to fool anyone. It is meant to be a joke. But it does illustrate in a bare-bones way all the basic elements of the direct *ad baculum*, put together in the right configuration to constitute a classic illustration of the structure of the fallacy. There is no doubt that the salesman is making a threat to Dagwood, and it is clear that he is using the threat as a method of moving Dagwood to a particular action. It is also clear that the salesman is doing this illicitly, instead of what he should be doing: persuading Dagwood to buy the product on the grounds that it is a good value for the money.

Argument from Consequences

A classic case of the indirect *argumentum ad baculum*, used by Copi (1986, 106) as an exercise for students to identify the fallacy, is based on argumentation from consequences.

Case 5.10

> According to R. Grunberger, author of *A Social History of the Third Reich*, published in Britain, the Nazis used to send the following notice to German readers who let their subscriptions lapse: "Our paper certainly deserves the support of every German. We shall continue to forward copies of it to you, and hope that you will not want to expose yourself to unfortunate consequences in the case of cancellation."

This case is a covert appeal to the threat of force or violence, given what is known about the capability and willingness of the Nazis to use force (of which the subscribers of this newspaper were presumably aware).

The case is an indirect *ad baculum* because, literally, the second conjunct of the last sentence is an instance of argumentation from consequences. Taken on the surface level, the "hope" that the reader does not want to "expose" himself to "unfortunate consequences" is a warning that alerts the reader to possible negative consequences of a course of action the reader may be considering. But of course, that's not all there is to it. At a deeper level of interpretation, the "warning" is really an appeal to the threat of violence.

Thus, the *argumentum ad baculum* here works by piggybacking on the argument from consequences. But it is not a species of argument from consequences. It uses argument from consequences to convey an indirect threat.

Argumentation from consequences is legitimate and very common. When two or more individuals are arguing about some proposed or contemplated course of action, it is appropriate for each of them to cite possible consequences of the action that this individual sees as relevant. It will be argued that positive consequences of the action support the case for going ahead with action and that negative consequences undermine or detract from support for the action.

A common context of dialogue for argumentation from consequences is advice-giving dialogue. For example, an expert who has been consulted may advise a layperson, "Don't go ahead with that course of action, because if you do, it will have highly negative consequences from your point of view." A financial expert might counsel, for example, that a particular action being considered is likely to create financial loss for the party considering.

Typically, the argument from positive consequences is pitted against the argument from negative consequences, one supporting one side, and one the other, of a disputed case. Much political debating about proposals and policies is based on this type of disputation. As noted in Chapter 3 (page 91), Windes and Hastings (1965, chapter 7) propose a model of political debate where the two opposed advocates dispute an issue, each arguing from the "positive" consequences of his or her own proposal against the "negative" consequences of the other side's proposal. Here "positive" and "negative" are defined by the presumed interests or point of view of a third-party respondent or audience.

Windes and Hastings (1965, 225–28) cite a kind of argumentation they called *argument from consequences*, where the conclusion expresses a "should" or "ought to" directive, an imperative expressing a policy or action. According to their account (227), one simple argumentation scheme of this type is the argument where an advocate lists and justifies the benefits or positive consequences that will result from a course of action being considered by a hearer. A critic can produce a counterargument by asking whether there could also be bad side effects (negative consequences) of the proposed course of action (226). Pro and contra argumentation then arises, pitting the case for positive consequences against the case for negative consequences.

Perelman and Olbrechts-Tyteca (1969) recognize argumentation from consequences as a legitimate and important type of argumentation common in everyday reasoning. Generally, they defined a *pragmatic* argument as one that "permits the evaluation of an act or an event in terms of its favorable or unfavorable consequences" (266). The pragmatic argument "regards the good consequences of a thesis as proof of its truth" (268). As an example, they cite the argument of the Stoic philosophers that happiness is the justification of their theories (268). In other words, the Stoics argued that people who embraced their philosophy were happy, therefore the Stoic philosophy was a good one because it produced beneficial results in practice.

According to van Eemeren and Grootendorst (1987, 289), the *argumentum ad consequentiam* consists of "testing the truth or acceptability of a standpoint by pointing out desirable or undesirable consequences." For example: "This can't be true because it would destroy everything this country stands for." They add, however, that whether an argument of this type is fallacious or not depends on the kind of proposition to be tested (295). If the proposition to be tested concerns a future course of action being considered, the *argumentum ad consequentiam* might not be fallacious at all. If the proposition is something to be tested in a scientific inquiry, pointing out its desirable or undesirable consequences may not be relevant to establishing it as true or false. In this latter context, arguing that a scientific hypothesis should be rejected because it has bad political or financial consequences, for example, could be fallacious.

As an example of fallacious argumentation from consequences, Fischer (1970, 301) cites the case where a reviewer condemned a psychoanalytic study of Martin Luther by writing, "In so far as it has been responsible for John Osborne's play, it may even need condemnation." Fischer comments that the critic was entitled to his opinion, but that the author of the study should not be condemned because of the "deficiencies of an unfortunate melodrama which may have been based upon it." Such cases demonstrate that the argument from consequences can in some instances be wrong-headed and rightly criticized as fallacious.

In other cases involving prudential or practical reasoning, argumentation from consequences is quite reasonable and appropriate. Practical reasoning is a species of goal-directed reasoning that seeks out a prudential line of conduct for an agent in particular circumstances.[4] Practical reasoning is oriented toward doing something, toward finding the best way, or at least an appropriate way, of doing something in a particular situation. By contrast, theoretical (discursive) reasoning seeks out evidence that counts for or against the truth of proposition.

Practical reasoning is oriented toward choosing a course of action on the basis of goals and knowledge of one's situation. The issue is to decide whether a course of action is practically reasonable or prudent in particular circumstances. Cognitive elements are involved in practical reasoning, but it has a prudential more than a cognitive orientation.

A practical inference is based on two characteristic premises. The one premise states that an agent has a particular goal or intention. The other premise states that there is a means for the agent to attain that goal. The conclusion of a practical inference states that the agent should carry out the action cited in the second premise. The conclusion expresses a practical "ought"; the agent ought to carry out the action in question on the assumption that he or she has adopted or is committed to the goal expressed in the premise.

Practical reasoning involves a chain of practical inferences where necessary and sufficient conditions for fulfilling goals are linked together so that the overall movement of the reasoning tends toward fulfillment or realization of the goal. Practical reasoning is practical because it occurs in a pragmatic context of argumentation.

One such context mentioned already is that of advice-giving dialogue, where an expert in a skill or domain of knowledge is consulted in order to solve a problem posed for a layperson who lacks direct access to the knowledge needed to solve the problem.[5] Another context of practical reasoning is that of planning. Planning arises when an agent considers

---

4. The account of practical reasoning given here is more fully developed in Walton 1990.
5. See the "*Argumentum Ad Verecundiam*" section of Chapter 5.

possible alternative future courses of action in attempting to fulfill a goal. Deliberation is a kind of planning that involves personal choices and conduct.

There are four kinds of critical questions appropriate for each use of a practical inference. The first critical question is whether there are alternative means of realizing the goal other than the action expressed in the given premise. The second critical question is whether it is in fact possible for the agent to carry out the action in question. If it is not possible, it may be best to change one's goals or to seek alternative actions that may be useful to fulfill one's goals. The third critical question asks whether the agent might have other goals that may conflict with the goal at issue. In a case of goal conflict, a decision should be made to change one's goals or perhaps to abandon one goal. The fourth critical question concerns possible or presumed consequences of carrying out an action. It asks whether there may be negative side effects that should be taken into account before a decision is made to go ahead with the action. Such a negative argument from consequences can be counterbalanced in pruden-tial discussions against alleged positive consequences said to be of greater weight.

In advice-giving dialogue, the argument from consequences often takes the form of a warning: "You ought to reconsider this proposed course of action because very bad consequences could ensue!" Within the context of a deliberation or planning dialogue, warning someone about negative consequences is very common and quite legitimate.

According to the analysis by Searle (1969, 67–68) of the speech act of "warning," there are three conditions necessary and sufficient for an act of warning.

1. *Preparatory Conditions:* (a) The hearer has reasons to believe the event in question will occur, and the occurrence of this event is not in the hearer's interest; (b) it is not obvious to either speaker or hearer that the event will occur.
2. *Sincerity Condition:* The speaker believes that the event is not in the hearer's best interests.
3. *Essential Condition:* What the speaker says counts as an undertaking to the effect that the event is not in the hearer's best interest.

According to a comment by Searle (1969, 67), warning is a speech act that is more like advising than requesting, because requesting counts as an attempt to get the hearer to perform an action. Warning, however, does not attempt to get the hearer to take action.[6]

Making a threat, by contrast, is not only an attempt by the speaker to get the hearer to carry out an action (or refrain from an action); it is an attempt to get the hearer to take evasive action in order to avoid the event cited in the threat.

Analysis of the indirect *ad baculum* exposes how an indirect speech act that appears on the surface to express a warning can be used to convey what is really a threat.

Three Levels of Analysis

Although many, including Keilkopf (1980) and Woods (1987), have categorized the *argumentum ad baculum* as a species of prudential or "prescriptive" argument that directs an agent to the conclusion that he or she ought (prudentially) to carry out a designated action, in fact the *ad baculum* is not this kind of argument. It is, however, closely related to prudential argument, and it is a major step toward the analysis of the *argumentum ad baculum* to see that it is characteristically based on the use of prudential argument. What is needed now is a more careful analysis of how prudential argumentation is used in the *argumentum ad baculum*.

The *ad baculum*, through the device of an indirect speech act, characteristically causes a shift from a warning in prudential argument to the making of a threat. On one level, a speech act of the type "Do X or some bad consequence B will occur!" can function as a legitimate part of a prudential argument in a context of dialogue where a proponent is warning or giving advice to a respondent. This imperative utterance functions as part of a sequence of practical reasoning, where the proponent addresses the respondent as "you" within the following Wreenish structure.

You were considering not doing X.

But if you don't do X, some consequence B, which will be very bad for you, will, or is likely to, occur.

Therefore, you ought to reconsider and (other things being equal) you ought (prudentially) to do X.

6. See the summary and alternative analysis given in Kreckel 1981 (45–50).

At this level, the argument is a warning, a kind of prudential argument that warns or cautions the respondent concerning the probable or possible negative consequences of a contemplated course of action (or inaction). But it is at the second, deeper level where the *ad baculum* argument arises. The warning utterance is placed in a context of dialogue where, as an indirect speech act, it properly entitles the respondent to draw the implication that she is being threatened by the proponent. At this second level of interpretation, the proponent is making a threat. He is, in effect, saying: "You had better do X, or if you don't, I'll see to it that B occurs." The original utterance, of course, does not literally say this. Literally, it is only a prudential argument—a warning or piece of advice. But secondarily, it functions as a threat in those cases where the argument is a genuine *argumentum ad baculum*.

Thus the *argumentum ad baculum* is not a species of prudential argument. It is an oblique tactic based on prudential argument in the first phase. It is a twisting of a prudential argument into a threat, through a secondary move that operates by suggestion and implication. It looks like a prudential argument on the surface, but it is meant to convey a threat, and that is where the *argumentum ad baculum* comes in.

It is true that the argument from consequences can be reasonable in some cases and fallacious in others. But it does not follow that the *ad baculum* is a special instance of the fallacious argument from consequences. This analysis of the *ad baculum* fallacy is simplistic, misleading, and essentially incorrect. The fallacious *ad baculum* argument is not a defective argument from consequences. It is a misuse of the argument from consequences in a complex argumentation technique that has two other stages. The second stage is the use of the argument from consequences (in an indirect speech act) to convey a threat. The third stage is the use of the threat to mask or substitute for a lack of evidence or good reasons to support a conclusion in argumentation or otherwise to go against or block legitimate goals of dialogue. All three stages (levels) of analysis are required in order to get an adequate reconstruction of the *ad baculum* fallacy as the distinctive type of fallacy portrayed by the textbook accounts.

At the first level, it is possible to have a fallacious argument from consequences, but this is not the same fallacy as that of the *ad baculum*. At the second level, a threat can be legitimate or reasonable, as well as illegitimate or failed in various ways. As Woods (1987, 344) points out, a bad prudential argument of the "stick-em-up" type could fail because, for example, it wrongly tries to exploit the interests targeted for violation by the proponent's threat. But, according to Woods, this is not a fallacy, but simply a factual error. Woods concludes that there is no *ad baculum* fallacy—that, in effect, the *ad baculum* fallacy does not exist. According to Woods's approach, arguments that appear to commit the *ad baculum* fallacy are really, when more carefully analyzed, threats that are in turn legitimate kinds of prudential arguments. Or, when they do go wrong, it is simply because they are weak, not fallacious, appeals to prudential considerations.

Woods's approach shows unusual insight into the failure of the superficial approach of the textbooks to pin down the *ad baculum* as a fallacy. Woods's argument is powerful and compelling—it shows that on the first two levels of analysis, no real *ad baculum* fallacy can be committed. There are only failed prudential arguments, failed threats and failed arguments from consequences. Woods skeptically concludes that there is no *argumentum ad baculum* fallacy. What is shown, however, is not that the *ad baculum* fallacy does not exist. What is shown is that there is no genuine *ad baculum* fallacy until one gets beyond this limited prudential framework.

As an example, consider the following case of the *ad baculum* used as an exercise by Copi (1986, 128).

Case 5.11

> Gentlemen, I am sure that if you think it over you will see that my suggestion has real merit. It is only a suggestion of course, and not an order. As I mentioned at our last conference, I am planning to reorganize the whole business. I still hope, however, that it will not be necessary to curtail the operations of your department.

In this case the speaker is presenting a plan for reorganization of a business at a meeting where the employees of one department are in the conference. The speech is addressed to these employees. The speaker is presenting a plan of action, but the dialogue is clearly a persuasion because the speaker's aim is to convince her audience that her proposal or "suggestion" has "real merit."

The speaker's last sentence is expressed as a "hope," but it can be interpreted in a literal way, at the first level of interpretation, as a use of argument from consequences. On this interpretation the speaker is pointing out that one possible consequence of the plan is to "curtail" the operations of the audience's department. This is cited as a possible negative consequence—the speaker "hopes" it will not be necessary.

At a second, deeper level of interpretation, however, the argument is not simply a citing of some negative consequences. In the context, it is quite clear that it will be taken (quite correctly) by the employees to be a threat. It is an indirect speech act used to convey a threat.

But a threat in itself is not a fallacious *ad baculum* argument. The threat, in this case, may be nasty and menacing, even unfair and malicious, but it does not immediately follow that it is fallacious.

The third level of analysis comes into play in this case when recalling the speaker's supposed original purpose—to persuade her audience of the "real merits" of her proposal. She is supposed to be gaining the support of those in her audience by giving good reasons or convincing evidence of the merits of her proposal. But instead, what does she do? She threatens members of the audience by suggesting, indirectly, that they had better support the proposal, or otherwise she will "curtail the operations" of their department. This sounds ominous—it is vague, but clearly very threatening. Perhaps, for example, it could involve terminating personnel in this department, that is, firing those people in the audience. Clearly the speaker's tactic involves an illicit shift—as a substitute for giving good reasons to support her plan, the speaker delivers an ominous threat. The thrust of this threat is clearly to close off any possible opposition and to stifle further critical discussion by the audience. Anyone who stands up to oppose the plan will single himself out for possible retribution. So the goals of the original critical discussion are effectively blocked by the speaker's *ad baculum* move.

Searle (1969, 67) distinguishes between categorical and hypothetical warnings, suggesting that most warnings are probably hypothetical. As an example of a hypothetical warning he gives the following statement form: "If you do not do X then Y will occur." Presumably, in the use of such a form as a warning, Y is an event or state of affairs that is bad for the hearer, and X is some state of affairs or event that the hearer can bring about or not.

Making a threat often takes this same form of hypothetical statement, but in this case, what is expressed is a different message. What is implied by the speaker is that if the hearer does not see to it that X is realized, then the speaker will personally see to it that Y does occur. The same statement form has been used, but what has been said is quite different, and is equivalent to the imperative "Do X or I will bring about Y!" The exclusive disjunction intended in the threat offers two choices: (1) do X and avoid the bad consequence Y, or (2) fail to do X and get the bad consequence Y.

But if the same form of statement can be used to convey a warning or a threat, what is the difference between the two speech acts? The two situations both require the potential for some bad event (for the hearer). The difference lies in the speaker's expressed commitment to intervene in the situation and bring about Y herself (if necessary) in order to get the hearer to do X.

Scaremongering and Intimidation

Threats and appeals to fear in argumentation are not necessarily fallacious tactics. But threats or appeals are likely suspicious, or even fallacious, in the context of a critical discussion, for threats are rarely appropriate to, or coherent with, the aims of such dialogue. In a political debate or any other context of dialogue that is like a critical discussion, using threats is a tactic that invites countertactics—the one who has used a threat (or is alleged to have used one) may be accused of using intimidation tactics.

While the *ad baculum* is sometimes a powerful tactic to use in argumentation, it is also a potentially dangerous one that may rebound onto the user. One who is perceived as using threats can be attacked as a brutal and vicious person who is not entering into the cooperative spirit of the critical discussion. Instead of playing fairly in the free marketplace of ideas and being "logical," such a person may be criticized as someone who is not open to rational persuasion, a kind of illogical "berserker."

Thus, the *ad baculum* is a tricky tactic to use in argumentation. It is highly effective in some cases, but in other cases it can backfire by inviting the deployment of powerful countertactics.

A fairly common criticism of an opponent's argumentation is that of *scaremongering*, that is, of appealing to the emotion of fear where such a response is not appropriate to the situation. The allegation is one of an unnecessary appeal to fear. In some cases, the scaremongering allegation is used to speed up a discussion by allaying fears and doubts concerning one's proposal, claiming that one's opponent's alarms are groundless. There is often a kind of personal attack in such a scaremongering allegation, saying in effect, "You don't need to pay serious attention to my opponent's arguments and delays, he is just trying to scare you; his tactics and motives are not honest or credible."

In the following sequence of debate (*Canada: House of Commons*

Debates, 18 August 1988, 1847) on the Free Trade Agreement (FTA), Mr. Sven Hovdebo argued that consideration of the agreement should be delayed until assurances could be given that existing government programs would not be threatened. Mr. Ross Belsher replied by accusing Mr. Hovdebo and his party of scaremongering; in so doing Mr. Belsher refers to the story of Chicken Little, who went around crying that the sky was falling in, trying to incite a fear that had no real basis in fact.

Case 5.12

Mr. Hovdebo: We have a clause which should be included because the Bill does protect regional development programs now and the need for them in the future. I notice that I am coming to the end of my time, Mr. Speaker, but there are two other motions both for greater certainty, both because the people of Canada do not trust this deal to protect them. These amendments give the kind of protection the people of Canada want. If this deal is to be forced on them, if the Government is to push through this deal without giving the people of Canada an opportunity to decide yes or no, to decide that they want it, this kind of security should be in the Bill. The social, environmental, self-government, and regional development programs which are in place or will be in place should be secured. We should have some security, ensuring that they will be in place after the Bill has been rammed through the House of Commons and has become law.

Mr. Ross Belsher (Fraser Valley East): Mr. Speaker, in listening to some of the speeches that have been made by the coalition on the other side of the House, I am reminded this morning of Chicken Little crying "The sky is falling." I guess that nursery rhyme is as true today as it was whenever it was written. We have people here who are not willing to look at what is really in the agreement or at what the agreement is all about. The FTA, as the Canadian public and all of us know, is an agreement we set out to put together with our neighbours to the south for the lowering of tariffs.

What we have now is both opposition Parties combining to try to put into legislation what is not in the agreement. We do not have anything in the agreement which does anything about our social programs, yet they are trying to scare Canadian people with misrepresentations about things that are not in the agreement. The legislation before us deals with incorporating into law what was agreed to with our neighbours to the south.

What Mr. Belsher was evidently trying to do was to speed up discussion of the FTA by counseling the House not to pay attention to the delaying tactics of Mr. Hovdebo and his party. He accused Mr. Hovdebo of emotional scaremongering, appealing to fears about issues that are not really related to or covered by, the agreement. His allegation was that Mr. Hovdebo and his party were blowing things up and trying to raise fears out of all proportion to what was justified. He argued that they were "not willing to look at what is really in the agreement," and were, instead, "trying to scare Canadian people with misrepresentations about things that are not in the agreement." The allegation was that the fears aroused by Mr. Hovdebo and his party were a purely emotional response, raised by them to scare the public, a response with no real basis in fact.

One can see the personal attack element surface in this use of the scaremongering charge when Mr. Belsher followed up his response (after a brief rejoinder about aboriginal land claims as affected by the FTA) by claiming that Mr. Hovdebo and his party were not telling the truth.

Case 5.12a

> Mr. Belsher: It really behooves the Canadian public to be careful what they listen to. I fear our Loyal Opposition is not telling the truth.
>
> Mr. Hovdebo: No more so than the Government.
>
> Mr. Belsher: They are not telling the people what is in the agreement. They are dealing with half-truths. Nothing is more dangerous than when people deal with half-truths and work up into such a lather that they start believing it themselves as being the truth.

According to Mr. Belsher, the FTA "is a good agreement, and it is time we got on with passing this legislation and putting it into effect" (1847). He concluded his speech by saying the opposition was being "pompous" in trying to stall this issue.

Here one side is trying to push the discussion ahead and the other side is trying to stall the move to legislation by arguing that there needs to be more discussion. Neither side has committed a fallacy, but one side has accused the other of appealing to fear and has exploited this tactic by building an *ad hominem* argument on it questioning the other side's motives and regard for truth.

In this case it would be unjustified and incorrect to claim that either side committed an *ad baculum* fallacy. But the "Chicken Little" move was an effective and interesting tactic.

The textbooks routinely use examples of *ad baculum* arguments that appear so repugnant or silly that their readers will accept them as fallacious without extensive justification or argument. However, in realistic cases, where there is a genuine difference of opinions on disputed issues, making such critical judgments may be nontrivial. In fact, what often happens is that when one side uses any argumentation that looks like it might involve some threatening appeal to force or fear, the other side immediately accuses them of using tactics of intimidation. Indeed, in some hotly disputed arguments, it is possible to find both sides claiming that the other is using tactics of intimidation.

Countercharges of intimidation were made when the liberal-minded administration of Dartmouth College suspended four students for "vexatious oral exchange, invasion of privacy, disorderly conduct, and harassment" stemming from their involvement in the conservative campus newspaper, *The Dartmouth Review*. The *Review* purports to present and advocate a conservative point of view and often attacks the political viewpoints and personal characteristics of selected professors. The dismissal arose specifically out of the *Review's* article on a music course taught by a professor whose use of "street language" in class was ridiculed. The article alleged that the academic standards in this class were low.[7]

After the *Review* sued the administration for violating its right of free speech, a report that interviewed key representatives of both sides of the dispute was televised on *60 Minutes*. The report began by indicating the conservative point of view of the *Review* and giving some examples of the kind of *ad hominem* argumentation used in it.

Case 5.13

Weekly, the *Review* attacks, sometimes brutally, Dartmouth's attempts to change its reputation of being all-white, all-preppie. The *Review* accuses the university of forsaking the intellectual tradition

7. CBS News, "Dartmouth vs. Dartmouth," produced by Norman Gorin, transcript of *60 Minutes*, vol. 21, no. 8, 13 November 1988, 7–10.

of the west, and routinely it savages selected faculty. Of a history professor: "Little Latin commies have found their patron saint of academia." Of a historian: "His courses are filled with fruits, butches and assorted scum of the radical left."[8]

One article was aimed at black students brought in under "affirmative action" and was perceived as "racist" by its critics, but a Dartmouth English professor defended the *Review*, saying: "The charge of racism as used against *The Dartmouth Review* is a slimeball attempt at intimidation that I think is absolutely reprehensible.[9] Note that the charge of intimidation is allied with an *ad hominem* attack through the use of the adjective "slimeball."

In reply, the president of Dartmouth College, James Freedman, made a countercharge of intimidation.

Case 5.13a

> I think I had an obligation to speak out against the *Review*, in part to protect the faculty, in part to protect what I believe are the important standards of Dartmouth College. I want to see uninhibited discourse here. I want to see discourse which is not intimidated by the fear of the kind of personal characterizations that appear in the *Review*.[10]

Here the countercharge is that the *ad hominem* attacks that have appeared in the *Review* are fear tactics used to intimidate uninhibited discourse. Towards the end of the dialogue the defenders of the *Review* once more made charges of intimidation when a former editor argued, "By kicking out the editor-in-chief and other top editors of this paper, the administration tried to intimidate and affect the editorial positions of this newspaper."[11]

This case is an extended dialogue where there is a genuine difference of opinion, and both sides appear to have a legitimate point of view. Or, at any rate, it is not obvious that the argument of one side is clearly fallacious or ridiculous. Yet both sides accused the other of using tactics of

8. Ibid., 7.
9. Ibid.
10. Ibid., 7–8.
11. Ibid., 10.

intimidation. Here it is no trivial question to judge fairly who, if anyone, committed the *ad baculum* fallacy of trying to block the dialogue by intimidating the other side.

In this case both sides were making a great show of being open to reasonable discussion in the tradition of academic fair play. Thus the charge of intimidation was a powerful one. In cases like this neither side wants to appear to use an *ad baculum* argument, and both sides are very keen to accuse the other of "intimidation," given the slightest excuse. Such accusations and counterclaims may be difficult to adjudicate, however.

Analysis of the Fallacy

In the following case one can see the effectiveness of the *ad baculum* tactic both in intimidating the opposition and in shutting down the continuation of a dialogue.

Case 5.14

> The trial of the terrorist group Direct Action was delayed in the French courts because of jury intimidation. The defendant threatened the jury with the "rigors of proletarian justice," on the first day of his trial, by asking, "I would like to know how long security measures will continue to be applied to the jurors?" Direct Action claimed responsibility for many recent terrorist attacks in France, and police suspected that the recent murder of a French auto executive was intended to frighten this jury. Evidently, the intimidation tactics were successful, for the trial had to be indefinitely delayed because so many jurors failed to appear in court.[12]

In this case, the threat was conveyed in an apparently innocent question about how long security measures would protect the jurors. But given the association of the speaker with the terrorist group Direct Action, the jurors clearly knew that the question was by no means an innocent request for information only. It functioned as an indirect speech act and was intended

12. Fred Coleman, "A Threat of Proletarian Justice," *Newsweek*, 22 December 1986, 38. This case is also discussed in Walton 1989a (99–100).

to be taken as a threat to the jurors, a threat meant to intimidate anyone on jury duty. It worked so well that the trial proceedings were indefinitely delayed.

If a fallacy is defined as the use of a tactic of argumentation to block legitimate goals of a dialogue, it is not hard to see how the *ad baculum* in case 5.14 fits this description. The intimidating question blocked off the sequence of dialogue that was supposed to take place during the regular trial procedure.

Ad baculum arguments can be classified into several different types. The central or paradigm type of *ad baculum* contains a threat, where a commitment to action is conveyed to the respondent by the proponent. It is also possible to have a secondary type of *ad baculum* argument that does not contain a threat but does contain an appeal to fear, or the citing of a threatening situation that appears menacing or ominous to the respondent.

TYPES OF *AD BACULUM* ARGUMENTS

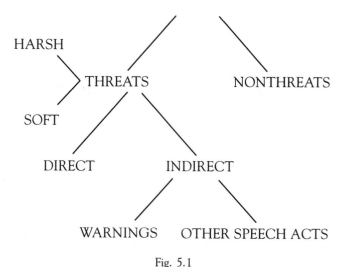

Fig. 5.1

The concept of threat in the above classification should be defined in a broad way. It should include not only *harsh* threats, those that appeal to force or violence, but also *soft* threats, those that appeal to sanctions not described as "force" or "violence." Soft threats are usually indirect, and they cite some consequences that the respondent would not like or that would not be in the respondent's interest; soft threats do not threaten the respondent with any physical harm or infliction of force. Harsh threats, by contrast, do essentially involve the application of force or the infliction of severe methods that would be physically painful or intimidating to the respondent. Soft threats are commonplace in bargaining and often involve financial losses or penalties.

A direct threat is an overtly expressed commitment to effect conse-quences that the respondent presumably will not like. There is no ambiguity in the expression of a direct threat. Indirect threats are expressed by indirect speech acts and work by suggestion (implicature). The proponent says one thing, but the respondent is invited to take a secondary message from it and draw the inference that a threat has been made (or an appeal to fear). According to Grice (1975) a *conversational implicature* is an inference in cooperative conversation where a hearer draws a conclusion from what a speaker has said, even though the speaker has not expressed that conclusion explicitly.

There are different kinds of indirectness involved in *ad baculum* arguments. The most common case is where the utterance is ostensibly a warning. It seems, on the surface, to have the form of an argument from consequences, but under the surface it is a threat. An example is case 5.10, the Nazi newspaper subscription case. The speech act is indirect. It is phrased as an argument from consequences. But it is ironic, and inten-tionally so. At a secondary level of interpretation, the speech act can rightly be taken as a harsh threat.

However, not every indirect *ad baculum* is a warning at the first level, which can also be taken as a threat. In the intimidation of a jury by the terrorist, case 5.14, the first-level speech act was a question. At a secondary level, however, the question functioned as a threat.

The first stage of analysis of the *ad baculum* argument then is to identify the speech act as a move that is being used as a threat. This is a non-trivial job of interpretation of a given text of discourse in relation to a context of dialogue. Especially in the cases of indirect threats, an extended analysis of the secondary meaning of the speech act may be required, an analysis that fits the evidence of discourse to the given particulars of the case.

The second stage is the evaluation of the *ad baculum* argument as fallacious or nonfallacious. This task requires an analysis of how the threat is used by the proponent in the dialogue. The first step is always to identify the type of dialogue that the participants are supposed to be engaged in. It may make a big difference whether it is supposed to be a critical discussion versus a negotiation, for example.

Once the type of dialogue has been identified, one can proceed to analyze how the threat is used by the proponent as a move or tactic in the dialogue. As has been shown, making a threat can, in some contexts, be a reasonable move (see case 5.7), especially soft and indirect threats. But if the use of the threat is illegitimate—in the sense that it has been used against the goals and accepted procedures of that type of dialogue—then a second step of the analysis has been carried through.

Finally, the third step is to analyze the *ad baculum*. How it is in fact used as a tactic of argumentation should be contrasted with how an obligation should properly be fulfilled in this type of dialogue. For example, the *ad baculum* fallacy could be a failure of relevance, used to mask the real issue or cover up a failure to fulfill burden of proof. Or it could be used as a tactic to shut down a critical discussion by intimidating one side.

What makes the *ad baculum* effective as a fallacy that really fools people in some cases is the existence or possibility of an underlying dialectical shift. Most often, what is involved is a subtle shift from persuasion dialogue to negotiation dialogue. The *ad baculum* can appear to be a plausible move in argumentation because it can be a legitimate move in negotiation dialogue. Using a threat of action or inaction to slow down or stall a negotiation dialogue is not necessarily an illegitimate move in itself.

Case 5.15

> In a meeting in Moscow with the U.S. Secretary of State, Soviet leaders put forward a "clear and formal" demand that the size of the reunified German Army be limited by a decision of the Two-plus-Four talks in Bonn. At these talks, the four major wartime allies are engaged in discussions with East and West Germany. The Russians wanted a ceiling to be set on the size of the German Army in the Two-plus-Four talks, and included in the East-West treaty on conventional arms being negotiated between NATO and members of the Warsaw Pact in Vienna. The Soviet Union announced that it was "freezing a unilateral withdrawal of troops from East Germany

pending progress on the Vienna talks and the solution to the German question."[13]

This last minute threat by the Soviets to freeze the withdrawal of their troops from East Germany would have the effect of slowing down or possibly even stalling the talks in Vienna and the Two-plus-Four talks. And there is a shifting in this case from one context of dialogue to another.

But in this case all three of the discussions involved were explicitly negotiation dialogues. There was no shift from a persuasion dialogue to a negotiation dialogue. Thus, although the threat could slow down or even close off dialogue, there is no basis for classifying the use of this threat as an *ad baculum* fallacy. For better or worse, such threats are a normal part of negotiation dialogue. They could be unreasonable, untimely, or unfair in various respects. But it need not follow that they are fallacious.

The Soviet demand could be classified as an *ad baculum* argument of a sort, on the basis that it does make a threat in order to get action or compliance to a demand. But there is no clear attempt to use the threat to persuade anyone that a particular proposition can be justified as true, or known to be true on the basis of evidence, or anything of that sort. So the argument should not be evaluated as an *ad baculum* fallacy. At any rate, it is not clear in this case that the requirements of burden of proof—for showing that the use of a threat constitutes a committing of the *ad baculum* fallacy—have been met. It is a kind of threat or appeal to force, but not an *ad baculum* fallacy.

The important thing is that not all *ad baculum* arguments are fallacious; one can't automatically presume that the use of a threat in argument is an *ad baculum* fallacy. There is a burden of proof on the critic who would make such a claim. And in the analysis given in this chapter, it has been shown how to go about fulfilling this burden.

A Pragmatic Analysis

The analysis of the *ad baculum* fallacy presented above is essentially pragmatic. It presumes a framework of a speaker and a hearer who are

13. Agence France—Presse Report, "Soviets Demand Treaty Limits on German Forces," *International Herald Tribune*, The Hague, 22 May 1990, 1.

partners in a dialogue that has goals and rules, and in a move of the dialogue, the speaker communicates a commitment to a type of action to the hearer, who is capable of recognizing this communication and reacting to it. This approach is in fundamental disagreement with Wreen (1988a), who thinks that *argumentum ad baculum* is not essentially dialectical— Wreen (91) claims that *ad baculum* is not necessarily a "two-person affair."

Characteristically, according to the analysis of the *argumentum ad baculum* advanced here, the context of dialogue requires a proponent who makes a threat and a respondent to whom the threat is made. And normally the proponent makes a threat that cites or exploits harmful consequences that will accrue to or happen to the respondent if he or she does not act in a particular way or accept a particular conclusion.

The *ad baculum* case presented below has a different twist. The proponent threatened to carry out actions that would bring harmful consequences to himself if his respondents didn't make a financial contribution by a deadline he set.

Case 5.16

> In 1987 Oral Roberts, the television evangelist, announced that he was 1.3 million dollars short of his goal of raising 8 million by a deadline, and that he planned to retreat to a prayer tower and fast until the money was contributed. Roberts wrote followers: "I'm going to be in and out of the prayer tower praying and fasting until victory comes or God calls me home." Alluding to the possibility that he could die, Roberts wrote, "If I go from there [the tower] to Jesus, I will see you in heaven." The purpose of the money was to send Oral Roberts University medical school graduates to be missionaries in Third World countries.[14]

According to the newspaper report, a Florida business man was "touched" by Roberts's appeal and signed a personal check for the required $1.3 million before the deadline.

Was Roberts threatening his followers, or was he really threatening himself? Certainly, the harm would have come to himself, but presumably his followers would also be harmed by being deprived of their leader.

Or perhaps Roberts was (although this sounds blasphemous) threatening

14. "Roberts Ransom Assured," *Winnipeg Free Press*, 22 March 1987, 1.

God to take action. In his letter, he wrote: "But I believe that [going from the tower to Jesus] won't happen, because I believe our God will do this mighty thing and at the end of March [the deadline], you and I will know the miracle has happened." On the other hand, from Roberts's point of view, no doubt it would seem inappropriate to accuse him of threatening God to take action. A defender of his point of view might deny that he was "threatening" anyone at all, saying that he was fasting and praying in order to achieve his goal.

The case is a curious one, however, for there does seem to be a threat of some sort involved, or at least an appeal or indirect threat to the possibility of harmful consequences. Yet instead of the usual case where one arguer threatens another, here the primary harm would appear to accrue to the proponent himself.

According to the analysis proposed here, the *ad baculum* fallacy is a tactic used in argumentation by a proponent in order to persuade a respondent (or carry out some other goal of dialogue) by making a threat or appeal to fear to that respondent. The tactic is fallacious where the appeal is used for the purpose of gaining the respondent's compliance or commitment to action in place of doing a proper job of giving evidence or proper argumentation that would support the goals of dialogue the proponent and respondent are supposed to be engaged in. This analysis is therefore in opposition to the account of the *ad baculum* given by Wreen (1988a). According to Wreen (91), in an *ad baculum* "the arguer needn't want to, or be trying to, move someone, or cause someone to do something." Wreen even goes so far as to say "an *ad baculum* can be offered with virtually any intention in mind, or even none, no overarching intention at all." To support this claim, Wreen (91–92) presents the following case.

Case 5.17

If one of the Twelve Disciples had said to Jesus, "Jesus, if you don't flee the city within the hour, you'll be arrested, tried, and probably crucified," he needn't want Jesus to leave town. He might simply be making sure that Jesus is apprized of his predicament, and/or making sure that Jesus is the man (deity?) he thinks he is. The argument might be offered, in other words, with no intention to move Jesus, and perhaps even in the hope that its conclusion, that Jesus ought to leave town, won't be acted upon. In fact, to get really

perverse about the matter, it might be offered with no definite hope
or intention in mind at all, but simply to see what Jesus will do; or
more far-fetched still, offered in a distrait, thoughtless manner, with
Jesus' disciple simply talking out loud to himself as he runs over
considerations in his own mind.

The analysis here derives quite a different conclusion from this case than
the one derived by Wreen.

Consider the first situation cited, where the disciple does not want Jesus
to leave town, but is only apprizing him of his predicament. On analysis,
this is not a threat but only a warning. It is in fact an argument from
negative consequences. The disciple is pointing out to Jesus that if he does
not leave, bad consequences are likely to occur. Surely there is no fallacy
in this.[15] It is simply an argument from consequences. It is not even an *ad
baculum* argument, much less a fallacious one.

In Wreen's second postulated situation, where the disciple is "simply
talking out loud to himself," there is even less of a basis for thinking there
could be an *ad baculum* fallacy. Assuming this "talking out loud" is not
really being used as an indirect way of conveying a threat, there is no *ad
baculum* argument here. Indeed, from all the information given in Wreen's
description of the case, there is no good reason to think that the disciple
is arguing at all. It is not an *ad baculum* fallacy. It is not an *ad baculum*
argument. It is not even an argument, so far as one can reasonably tell. In
these cases, what Wreen cites as examples of *ad baculum* do not even count
as *ad baculum* arguments by the criteria of this book's analysis.

It is a thesis maintained here that an *ad baculum* threat succeeds only if
the proponent's expressed intention to cause harmful consequences to the
respondent is recognized as such by the respondent. Thus, an *ad baculum*
argument requires not only a dialectical framework containing a proponent
and a respondent; it also requires that the proponent express a particular
intention and that the respondent recognize and understand this intention.
Analyzing the *argumentum ad baculum* goes deeper than a semantic analysis
of a set of true or false propositions. It is a pragmatic concept relating to the
expressed commitment of a speaker who carries out a speech act and to
the effect this speech act is designed to have on a respondent with whom
the speaker is engaged in dialogue.

15. There is no intention to give the impression here that Wreen thinks the *ad baculum* is a
fallacy, for he appears to be generally skeptical of the claim that the *argumentum ad baculum* is a
fallacy.

The *ad baculum* is inherently pragmatic in another way as well. The indirect *ad baculum* argument expresses a warning, on the surface, but also works through the exploitation of an ambiguity. At a secondary level, beneath the surface dialogue, it is meant as a threat. Whether it is reasonably taken as a threat, however, depends on the particular context of dialogue in a given case. What is an innocent warning in one situation could rightly be taken to be a menacing threat in another situation. Much depends on how the speech act is expressed.

The indirect *ad baculum* argument is based on an indirect speech act—a speech act that, used or taken literally, means one thing, while used or taken in another way, means something different. The classic example of an indirect speech act is "Can you pass the salt?" which can be used literally as a question but standardly functions as a request. Although the indirect use, in this case, is more common than the direct use, as Bach and Harnish (1979, 175) pointed out, features of the situation strongly influence how this sentence would normally be taken. If uttered at the dinner table, it would normally be interpreted as a request to pass the salt. If uttered to a patient recovering from polio, by a physical therapist, however, it could be interpreted as a question. To interpret an indirect speech act, therefore, one has to take a close look at the context of dialogue in which it is uttered.

The same is true of the warning-based indirect *ad baculum*. What could normally be taken as a warning in one context of dialogue could quite reasonably be interpreted as a threat in another context. It follows that there is a special problem of pinning down the indirect *ad baculum* as a fallacy in particular cases. It is a problem of communication.

According to van Eemeren and Grootendorst (1990), when language users perform communicative speech acts, they should observe certain rules for successful communication. One of these rules is to "be clear," that is, to formulate the speech act in such a way that the respondent can recognize what proposition is expressed in it and what the communicative force of the speech act is meant to be. This does not rule out indirect speech acts, but it does mean that the proponent of an indirect speech act is under no obligation to make clear what his speech act means.

However, with the indirect *ad baculum*, it is characteristically the case that the proponent has expressed the threat in an indirect speech act intentionally, in order to escape the charge that he or she has made a threat. In many cases uttering an explicit threat would be a violation of the accepted rules of discourse. Making a threat could be impolite or even illegal. In a union negotiation or a parliamentary debate, for example, threats may simply not be allowed. What might be allowed, however, are "warnings" that could be taken as threats, given enough plausible deniability should an accusation of threatening or intimidation be raised.

Clearly, in many cases, the tactic of the indirect *ad baculum* is to avoid the appearance of commitment to a threat. The tactic avoids explicit commitment while at the same time getting across a message to the respondent that will make a mark, leaving the respondent in no doubt of the proponent's real commitment (unexpressed intention).

With the indirect *ad baculum*, even before one gets to the stage of evaluating an instance of it as fallacious or nonfallacious, there arises the problem of interpretation. Can it be pinned down as a genuine threat? The problem here is to interpret the given discourse in relation to the rules of communication and, in particular, in relation to the rule of clarity. Indirect *ad baculum* arguments being what they are—often subtle and even ironic—it is often possible to suspect the proponent of equivocation, even before considering the question of whether the threat is a substitute for a legitimate argument. And prior to that is the question whether the speech act is really a threat, or whether it can correctly be so taken in the context of dialogue.

Although it is clear that the notion of a threat is important in understanding how the *ad baculum* works as a fallacy, nevertheless Wreen is correct in saying that threatening is not essential to the *argumentum ad baculum*. In an *argumentum ad baculum* case the argument might work by appealing to fear, through the use of tactics of intimidation or scaremongering. Generally, the *argumentum ad baculum* is best defined as the type of argument that derives its weight of presumption in a dialogue through the use by one participant of an appeal directed to the fear or timidity of the other participant. So defined, this type of argument is characterized as an appeal by one party to the emotions, or the susceptibility to emotion, of another party in a dialogue.

However it also needs to be recognized that such an appeal may, and often does, contain the making of a threat, either overtly or covertly, by the first party. Although an act of threatening is not absolutely essential for an argument to be *ad baculum*, in fact the majority of the most interesting *ad baculum* arguments involve threats. In the covert cases, the task of evaluation is to judge whether a threat is really made or not.

The job of interpretation, analysis, and evaluation of *ad baculum* arguments turns out to be a pragmatic and linguistic task that requires careful attention to the language and context of dialogue of a particular case. For this reason the study of the *ad baculum* is well suited to the case study approach, which looks at each individual case on its merits. The job requires the identification of the participants' commitments in a dialogue, based on the evidence of what they say and how they react to each other's speech acts.

The occurrence of a threat in a dialogue is often rightly to be taken as a sign or warning indicator of an *ad baculum* argument. But it is not an infallible sign of the committing of an *ad baculum* fallacy. To back up a charge of fallacy, one must go through the proper steps of interpretation, analysis, and evaluation outlined in this chapter, in order to meet the requirements of burden of proof.

6

ARGUMENTUM AD HOMINEM

The conventional wisdom recognizes four basic subtypes or classifications of the *ad hominem* argument, or argument against the person. In the *abusive ad hominem*, an arguer's character is attacked in order to discredit her argument. In the *circumstantial ad hominem*, an arguer's personal circumstances are said to conflict with her argument, once again with the object of discrediting her argument. In the *poisoning the well* variant of the *ad hominem* argument, an arguer is said to have shown no regard for the truth, the implication being that nothing she might say henceforth can be trusted as reliable or sincere. In the *tu quoque* type of *ad hominem* argument, a participant in the argument replies to his opponent's criticism by saying, "You're just as bad yourself," the implication being that she has no right to make this criticism, and her making it can be discredited or ignored.[1]

---

1. For accounts of the conventional treatments in logic textbooks of the *argumentum ad hominem* as a fallacy, see Govier 1983 and Walton 1985.

This chapter introduces a new taxonomy of *ad hominem* arguments that departs from the conventional system of classification in several respects. First, a new type is introduced. In the *bias ad hominem* argument, one arguer claims that the other has something to gain, a "hidden agenda," and that therefore her argument should be discredited or discounted as deceptive and unreliable. Second, the poisoning the well variant is proposed to be a subtype, an extension of the bias type of *ad hominem* argument. Third, it is proposed that the abusive classification should be rethought and renamed the *personal ad hominem argument*. Fourth, it is argued that the *tu quoque* should not be regarded as a basic category or type of *ad hominem* argument on a level with the other types. Instead, it is suggested that the *tu quoque* is an argument tactic in its own right, but one that is often and quite effectively added on to any one of the other *ad hominem* argument types as a way of extending it or counterattacking against it.

In defining the *argumentum ad hominem* it must be decided whether any personal attack against an individual counts, or whether the personal attack has to be used to denigrate or rebut this individual's argument. This is addressed in the last section of this chapter.

The introduction of an *ad hominem* argument into a dispute represents the personalization of the dialogue. Quite expectedly and characteristically, therefore, the use of the *ad hominem* leads both to an intensifying of personal involvement in a discussion and to a heightening of emotions. Indeed, one of the main problems of using an *ad hominem* argument is that it has a way of making a discussion into a quarrel. Once the *ad hominem* is used, it is often used by the attacked person again, in *tu quoque* fashion, and as personalization of the argument deepens, it becomes difficult to turn back to a less personalized discussion of the issue. In fact, too often what happens is that the original critical discussion deteriorates into an eristic dialogue or species of "verbal combat" where critical restraint is left behind, and the only aim is to "hit out" verbally to injure the other party.

Despite the dangers of personalizing an argument, this chapter will deny that the *argumentum ad hominem* is a "raging beast" that always runs out of control, destroying a reasonable discussion. Perhaps unexpectedly, this chapter will argue not only that personalization may be helpful to a critical discussion in many cases but also that it is the most important single benefit of a successful critical discussion.

Are *Ad Hominem* Arguments Necessarily Fallacious?

The textbooks often presume that the *ad hominem* argument is always fallacious. The idea is that this type of argument is so reprehensible and underhanded that it must be fallacious.

However, a little reflection will show that such a position is not tenable. In the case below, for example, if the personal allegations are based on good evidence, the *ad hominem* argument is reasonable. Suppose Smith has put his name forward as a contender for the position of a magistrate, but the following argument is voiced against his contention that he would be a good person for the position.

Case 6.1

> Smith is not a good candidate for the position of magistrate because he is a dishonest and violent person who has been convicted several times for larceny and assault and is a member of the American Nazi party.

In this argument, one should question the truth of these personal allegations, but it may be hard to deny their relevance to the issue of Smith's suitability as a magistrate if the allegations are justified by evidence. The lesson is that in some cases personal characteristics are relevant to an argument. In case 6.1 the position of magistrate is rightly seen as requiring a person of good character and good judgment. Going into a candidate's personal history and moral character is therefore a legitimate and relevant consideration.

It used to be that the press followed a convention of politeness in respecting a political officeholder's right to privacy by not asking questions regarding the person's ethical conduct, especially in regard to sexual matters. But recently this trend has changed, as more and more ethical questions came to be perceived as legitimate. Generally, it is now held that the "character issue" is a legitimate subject of reporting and political discussion.

Actually, many aspects of character have always been considered relevant in political campaigning, debating, and reporting. But what was illustrated by a series of cases, starting with the notorious Gary Hart case, was that family and sexual matters that were formally considered "private" by conventions of polite discussion now came to be considered legitimate and proper issues for political discussion.[2] This focus on the personal precipitated an "ethics wave," where ethics committees began to scrutinize carefully the personal and financial connections of leading politicians.[3]

Case 6.2

> Speaker of the House Jim Wright was charged with violations of House of Congress ethics rules: using the sales of his book, *Reflections of a Public Man*, to evade limits on speaking fees, using his influence to get his wife a "no-show" job, free use of a condominium and a Cadillac, and pushing through legislation that his Texas business associate had a direct interest in.[4]

This "ethics wave" went so far that some critics spoke of "pious invocation" of the "character issue" as an excuse to assault an opponent by "mudslinging."[5]

The textbooks are right to emphasize the fallacious aspects of using *ad hominem* arguments. Quite clearly it is a powerful and dangerous form of attack in argumentation that can easily get out of hand, causing much mischief and confusion, even making a critical discussion go "off the rails" altogether. On the other hand, there is now a well-established line of argument in the literature on *ad hominem* that defends it as a type of argumentation that can be reasonable in some cases, when used under the proper conditions.

John Locke described the *argumentum ad hominem* as pressing "a man with consequences drawn from his own principles or concessions" (quoted in Hamblin 1970, 160). Locke did not describe this type of argumentation as a "fallacy," however, and Johnstone (1978) followed up Locke's neutral account by arguing that *ad hominem* arguments from an opponent's concessions can be quite reasonable and constructive in many cases.

Brinton (1985) argues that the *argumentum ad hominem* is a legitimate strategy of rhetoric in public deliberation, and ought not to be regarded as

2. Jonathan Alter, "Sex and the Presidency," *Newsweek*, 4 May 1987, 26.
3. Tom Morganthau, Howard Fineman, and Eleanor Clift, "Ethics Wars: Frenzy on the Hill," *Newsweek*, 12 June 1989, 14–18.
4. Ibid, 14.
5. Tom Morganthau and Eleanor Clift, "Looking for an Exit," *Newsweek*, 5 June 1989, 18–20.

a term of abuse. Barth and Martens (1977) analyze the *ad hominem* fallacy as the following type of argument in a formal game of dialogue: a thesis has been defended successfully on the basis of one's opponent's concessions in a dialogue; therefore that thesis is absolutely correct (defensible against any opponent who makes any concessions he or she might care to choose). But Barth and Martens presume that *ad hominem* argumentation (arguing from an opponent's concessions in dialogue) is not inherently fallacious and that in many cases it can be quite reasonable and legitimate as a kind of argumentation.[6]

Despite its portrayal as a fallacy by logic textbooks, the *ad hominem* has been recognized as a reasonable type of argument elsewhere. According to legal rules of evidence, it can be legitimate to question or criticize the character of someone testifying in court and, in particular, to attack the person's character for veracity. One context of argument of this sort is the interrogation of a witness. The best known case is the cross-examination of witnesses in the law courts. Here impeachment, the process of showing facts that reflect on the veracity of a witness, is recognized as a legitimate kind of argumentation that can lead to the rejection of that witness's testimony. Legal standards of evidence recognize legitimate as well as illegitimate forms of criticism of the character of a witness in the process of impeachment.

Generally, character can be a legitimate and relevant factor in this context of argumentation because it may be difficult to get at the facts of a matter, and the circumstantial evidence in a particular case may be indecisive. Where there is insufficient direct or decisive empirical evidence to adjudicate on a case, it may be correctly regarded as legitimate to go by the say-so of eyewitnesses who are not experts but whose opinions are taken on trust that they are reliable and trustworthy witnesses. However, if evidence that they may not be trustworthy arises in the course of their cross-examination, this is a good reason for rejecting the plausibility of their testimony. Evidence may be found, for example, that the witness has lied in a previous case, or that the witness has contradicted herself or himself (see Loftus 1979). This criticism is *ad hominem*, and yet, in some cases it may be perfectly legitimate as a way to evaluate the worth of the testimony of a witness, and this kind of argumentation has been accepted in principle and subject to qualifications as reasonable and legitimate in law courts.

6. An extensive account of the literature on the *ad hominem* is given in Walton 1985.

For example, Degnan, in his article on evidence in the *Encyclopedia Britannica* (1973, 908), cites four universally recognized forms of impeachment. The first type of impeachment is showing that the witness has previously made statements inconsistent with those now made under oath. The second type of impeachment is showing that the witness is biased, either for or against one of the litigants. The third type of impeachment is showing general bad moral character of the witness. And the fourth type of impeachment is showing that the witness's character for truth and veracity is bad. This first type of impeachment is actually a species of circumstantial *ad hominem* argumentation. The second type of impeachment is what is called above a biased type of *ad hominem* argument. And the third and fourth types of impeachment coincide with the personal or abusive type of *ad hominem* argument.

These three types of *ad hominem* argument have been cited in works on legal evidence as reasonable arguments, reasonable at least as criticisms that can be used to discredit the testimony of a witness in the context of cross-examination in a legal trial.

On the other hand, some cases of *ad hominem* argumentation have been recognized in the law of evidence as illegitimate or incorrect types of argumentation. For example, in a previous article on evidence in the *Encyclopedia Britannica* (1960, 16), Sir Courtenay Ilbert wrote that evidence is not admissible if it is meant to show that a person who is alleged to have done something was in a disposition or character that makes it probable that he would have or would not have done it. Thus, legal tradition and standards of evidence recognize acceptable as well as unacceptable forms of *ad hominem* argumentation.

Source-Based Evidence

Why should the character of an arguer be a legitimate or relevant consideration in assessing his or her argument? There are two reasons.

First, the argument may itself be about character. For example, in political campaign argumentation, the character of the candidate is, at least partly, what the argument is about. By contrast, in a scientific inquiry, for example, in physics or mathematics, the argument is not about the character of the scientists, and any *ad hominem* argument would (rightly) be discounted immediately as not a legitimate consideration to play any part in a proof.

Second, when objective evidence is lacking, it may become necessary to rely on testimonial evidence, and thus on the character of witnesses. When objective evidence is insufficient to determine a conclusion and, for practical reasons, a conclusion of some sort, even a tentative conclusion, is prudentially reasonable under urgent circumstances, presumptive argumentation comes into play. In such cases, the argument may have to rely on various kinds of testimonial evidence like eyewitness testimony or expert opinions. The trouble with these source-based appeals is that they are founded on trust, and the source may be lying. However, the reliability of sources can be examined, they can be made to take oaths to be truthful, and their testimony can be "tested" in various ways. The way to proceed is to presume that a source is honest and sincere, unless evidence or indications to the contrary turn up.

The *argumentum ad hominem* is a way of verifying the credibility of a source in presumptive reasoning. There are two kinds of evidence or justification that can be used to back up an assertion that is challenged in argumentation.[7] First, one can check source-independent evidence for or against a proposition. Ideally, source-independent evidence is knowledge-based reasoning. Second, one can evaluate the weight of presumption for or against a proposition by checking the source. This source-dependent evidence is not directly based on knowledge but is based on presumptions about the access of the source to knowledge and the credibility of the source in presenting and making available the knowledge to a user. The credibility of the source may be identified with what Aristotle called *ethos*, and the type of argumentation associated with it is described as ethotic argumentation by Brinton (1986).

In evaluating the weight of presumption for or against a proposition by checking the source, there are six critical questions.

1. Is the source consistent? The previous explicit commitments or pronouncements of the source need to be checked against what he is saying now. If he is "changing his tune," can he give reasons to explain the inconsistency?
2. Is the source honest? Here the question is whether the source has a good character for veracity. Also, it can be asked whether, in this case, the

7. See "*Argumentum Ad Verecundiam*," Chapter 5, on hard and soft evidence.

source has any reasons for lying, or being less than honest or frank in offering an opinion.

3. Is the source sincere? Is what the source says in this case consistent with what can be inferred presumptively from the information given on his position generally? If not, can he give reasons to explain the variance? Here the question is one of consistency between the opinion given by source and his nonexplicit commitments to his deeper position on the issue (as far as this is known).

4. Is the source reliable? Can the source be depended upon to back up his expressed opinions by producing good evidence and explanations? Is he serious and thoughtful in expressing opinions? Will he retract a commitment if new information or evidence becomes available or is cited?

5. Is the source of good moral character? Above the question of honesty, does the source have a history of cooperative and social behavior? If there are cases where the moral character of the source has been doubted, can these be defended?

6. Does the source have good judgment? Does the source have the skills of good judgment required to formulate a properly qualified opinion? Can the source deal with problems of contingencies of a particular case and exceptions to a rule in an effective and balanced way?

In evaluating source-dependent evidence, asking any of these six questions is a way of challenging the presumption that the say-so of the source should carry evidential weight. Unless the question can be answered satisfactorily, the presumption in favor of the proposition vouched for by the source is defeated. Hence, source-based reasoning is inherently defeasible. It goes ahead on the basis of a provisional weight of presumption that is subject to rebuttal if the credibility of the source can be effectively challenged by cross-questioning.

Attacking a person's argument in cross-questioning by challenging the presumption that this person has a good character, is honest or sincere, can be a reasonable kind of argumentation. It could be called the personal *ad hominem* argument. This type of *ad hominem* argument is most often called the "abusive" *ad hominem* in the textbooks, but this term suggests that it is inherently wrong or fallacious (Govier 1983; Walton 1985). For this reason, it is better to call the generic type of argument the personal *ad hominem* argument and reserve the "abusive" label for the fallacious instances of its use.

If the personal *ad hominem* argument is sometimes nonfallacious, it is also true this type of argumentation is subject to flagrant and notorious abuses. Quite often, personal attack turns out to be based on slander, innuendo, unsubstantiated rumor, or false allegations of personal misconduct. In fact, such charges are often fabricated precisely for the purposes of attacking a person's argument, and the use of this tactic has often enough proved itself to be powerful and successful in convincing an audience not to accept the argument.

What then is the difference between the reasonable and the fallacious use of the personal *ad hominem* argument? In making this evaluation, three factors should be kept in mind.

First, can the charge of bad character or misconduct alleged be backed up by good evidence? This is the question of whether the premise is right or justifiable.

Second, what is inferred from this premise? Does the *ad hominem* attack merely raise critical questions about the reliability of the source, or does it conclude (absolutely) that what the source said must be false? The distinction here is between strong refutation (absolute rejection) and weak refutation (raising critical doubts concerning a presumption). Weak refutation leaves open the possibility of reply and rebuttal.

Third, the context of dialogue is extremely important. In a scientific inquiry a personal *ad hominem* attack might stand out as inappropriate and fallacious, whereas in a hotly contested trial where witnesses are divided and opposed, *ad hominem* arguments could be quite legitimate and even decisive in shifting a burden of proof to one side or the other of a case.

Ethos: The Pedestal Effect

Political argumentation is a context of dialogue where perceived moral character is, and should be, a strong factor in determining a speaker's credibility before an audience. Aristotle's theory of rhetoric can be reconstructed in such a way that it systematically yields a view of the *argumentum ad hominem* as a reasonable type of argumentation in political discourse (Brinton 1985). According to Aristotle's theory of practical reasoning, political discourse is concerned with deliberations that try to arrive at a right or prudent course of action. Such decisions are about things that are "variable," however, and the questions raised do not admit of answers that are "knowable" on the basis of scientific calculation. Accordingly, a prudent and virtuous person who has had a long experience of successfully dealing with deliberation can rightly have great influence on public decision making as an orator. But the assumption is that this person, in order to be influential and have the people take his or her advice seriously, must be (or at least appear to be) a person of good character (*ethos*).

This sketch of Aristotle's theory of political rhetoric is the background structure required to support his concept of what Brinton calls *ethotic argument*, "a kind of argument or technique in argument in which *ethos* (character) is invoked, attended to, or represented in such a way as to lend credibility or detract credibility from conclusions which are being drawn" (1986, 246). Within this Aristotelian framework of political deliberation, Brinton sees ethotic argumentation based on a speaker's character as, in principle, a legitimate and reasonable kind of argumentation.

Aristotle's proof of the importance of character in political argumentation can be broken down into five steps. The basic framework of political speechmaking involves one person (a speaker) influencing many people. Thus, this framework is applicable where there is a leader, hero, or spokesperson addressing a large group. The first step is to identify the type of speech event as a deliberation where the group is trying to arrive at a prudent decision on a course of action. This type of discourse then involves policymaking or a decision concerning a future course of action. It is a situation for practical reasoning.

The second step is the thesis that practical reasoning in a "real-world" situation inevitably involves uncertainty, due to the complex nature of variable circumstances, especially in politics. Thus, the conclusion must be arrived at on the basis of presumption rather than knowledge. It is not a question where a scientific inquiry would be the appropriate method to solve the problem (by itself).

Therefore, the third step is that the advice of some wise, experienced person could be relevant. Since the required argumentation outruns knowledge or scientific calculation, an alternative source of guidance could be to turn to the counsel of someone who has had experience in dealing with similar situations and has good, practiced skills of judgment and sizing up this type of situation.

The fourth step is that turning to the advice of an authority presumes that the sincerity, honesty, and integrity of this person can be trusted. The wise person could be lying or might be concealing some flaw of character or judgment. It is a Gricean presumption that the person in question is committed to requirements of honesty and sincerity in dialogue.

Therefore, the fifth step, and conclusion, is that character is important in political deliberation dialogue where a speaker is influencing the decisions for action of a group he or she is addressing. Especially important is character for veracity, sincerity, and good practical judgment.

These five steps can be summarized as follows.

1. Speech event of deliberation—practical reasoning.
2. Variable circumstances produce uncertainty.
3. Experienced counsel of the wise could be relevant.
4. Presumption of honesty, sincerity, and judgment skills.
5. Therefore, the character of the speaker is important.

Aristotle's proof postulates a normative model of political debate as an ideal of reasonable dialogue. Some might question the applicability of this model to modern democratic political discourse, especially at step five. But assuming that Aristotle's five steps do constitute a model of some sort of ideally rational political dialogue, it follows that the *argumentum ad hominem* has an important place in political debate as a kind of argumentation that is, at least in its rightful context, occasionally reasonable and legitimate.

The problem, however, is that political debate is a complex type of dialogue that contains elements of the eristic dialogue as well as elements of the critical discussion. Bringing in the character of an arguer can have side effects that can lead to an abandonment of critical discussion and a shift to personal quarreling. Evaluation of the *argumentum ad hominem* as fallacious or reasonable, therefore, may turn on a razor's edge. In principle, attacking your opponent's character in a political debate can be reasonable and legitimate, actually contributing to the quality of the debate. But in practice, an *ad hominem* attack on character can often go badly wrong, interfering with or even blocking the debate altogether.

Circumstantial *Ad Hominem*

The second basic type of *ad hominem* argument is the circumstantial *ad hominem* argument, a questioning of an arguer's position by citing a presumptive inconsistency within that position. Typically, the inconsistency alleged is a practical inconsistency rather than a purely logical inconsistency. For example, the person's principles that she has advocated in her argument may be alleged to be inconsistent with her known personal actions of the past. The term "circumstantial" is appropriate because the inconsistency relates to the personal circumstances of the individual in relation to the propositions she has advocated in her argument. Hence, this type of criticism can be identified with the expression "You don't practice what you preach." The difference between the personal *ad hominem* argument and the circumstantial *ad hominem* argument is that the circumstantial attack involves a criticism of inconsistency. By contrast, in the abusive type of argumentation, no inconsistency is directly involved. Instead, the argument is a direct attack on the individual's character.

The circumstantial *ad hominem* argument cites some personal circumstances of an arguer that appear to place her in a position where she contravenes her own argument. Needless to say, this often makes an arguer look very bad. It can make an arguer look insincere, hypocritical, or even thoughtless or foolish, thereby undermining her credibility and her argument at the same time. This powerful criticism, in effect, suggests that this person really has no right to put forward such an argument criticizing others when he himself is guilty of the same alleged fault.

In a *Newsweek* interview, Lee Iacocca, chairman of Chrysler Corporation, reacted to the numerous criticisms of American attitudes that had recently been expressed by the Japanese. Among the many criticisms replied to was the following.

Case 6.3

> On Japanese criticism of American 'greed': [Sony chairman Akio] Morita says [U.S.] management is too fat, they're too high paid, here's a guy who's making billions of dollars a year and worrying about his tax bill. ("One Hell of a Train Wreck," *Newsweek*, 2 April 1990, 21)

The implied conclusion is that Morita's criticism can be discounted and rejected, because he is in the same position (or better) as those he criticizes for being too highly paid. The circumstantial *ad hominem* attack undercuts the credibility of the speaker.

In a classic case of the circumstantial *ad hominem* argument, a child criticizes a parent's argument against smoking by citing the parent's own personal circumstances.

Case 6.4

> Parent: There is strong evidence of a link between smoking and chronic obstructive lung disease. Smoking is also associated with many other serious disorders. Smoking is unhealthy. So you should not smoke.
>
> Child: But you smoke yourself. So much for your argument against smoking. (Walton 1989b, 152)

Looking at the parent's argument in an objective way, it could be a good argument, based on good medical evidence that supports a prudential conclusion that one should not smoke. If the child is dismissing this evidence *tout court*, she could be committing an *ad hominem* fallacy.

But you can also look at the child's argument in a subjective way. The child is not in a position to understand or evaluate this medical evidence directly, so the best she can do is to take her parent's word for it. But the parent smokes. If so, the child may be reasoning, how can the parent be a serious or sincere advocate of nonsmoking? Interpreting the argumentation in this way, it seems that the child could be raising a legitimate question.

So is the child's *ad hominem* argument against the parent fallacious or not? It depends on whether the child's argument is a strong or weak refutation of the parent's argument. If interpreted as a weak refutation, a kind of critical questioning that raises doubts about the parent's sincerity, the child's *ad hominem* reply is reasonable enough. However, if the child's reply is meant as a strong refutation that wholly dismisses the parent's argument, the reply is an instance of the *ad hominem* fallacy. Just because the arguer exhibits a personal or circumstantial inconsistency in advocating his or her argument, it does not follow that the argument, in itself, is a bad one, or that the conclusion is false. So to presume would be to commit the circumstantial *ad hominem* fallacy.

So should the child's response be interpreted as committing her to a strong or only a weak refutation? The textual evidence of the dialogue must be scrutinized carefully. On the whole, this evidence suggests that the child is making a strong refutation move, dismissing the parent's argument altogether by saying, "So much for your argument against smoking." If so, the child has committed a circumstantial *ad hominem* fallacy.

Naturally, however, this charge of fallacy is a conditional one, depending on a particular interpretation of what the child has said in the dialogue.[8] Such a charge could be withdrawn if the child were suitably to qualify or clarify her remark. The important point is that the inference from circumstantial inconsistency of an arguer's position to the absolute rejection of his or her argument is a logical leap. To make the leap too hastily or uncritically is to commit the *ad hominem* fallacy.

The effect of the circumstantial *ad hominem* in destroying the credibility of someone who has been "put on a pedestal" is illustrated by the following case. It was reported that Morten Harket, Norway's top pop singer and teenage idol had offended his fans and compatriots by buying a new Mercedes 500 SL car. Harket, an advocate of the environmental cause, had publicly endorsed the "green" position by advocating that people should drive electric cars in order to save fuel and cut down on pollution. However, his car was a large Mercedes that used a lot of gas.

Case 6.5

> "Look, this is the environment boy's petrol-guzzler," said the front-page headline in the newspaper Dagbladet, after a smiling Harket featured in several newspapers today, posing next to his new Mercedes 500 SL.[9]

Harket was reported to reply: "I need a car and I'm lucky enough to be able to afford this one." It would seem likely, however, that this response did nothing to reassure his disappointed fans.

Why was this response inadequate? A teenage idol, a kind of hero or "role model" for young people, has allied himself with a popular moral cause. But his personal actions appear to be going contrary to the policies he is recommending for others. Such apparent hypocrisy destroys his credibility as a leader or role model figure, for it suggests that he is insincere, that he is not personally committed to the policies he advocates.

In case 6.5 the *ad hominem* criticism pointing out the circumstantial inconsistency is reasonable enough provided it is taken as a questioning of the arguer's sincerity in advocating the environmental cause. As case 6.4

---

8. See also the further discussion of this case in Walton 1985 and 1989b.

9. Reuter report from Oslo, Norway, "Rock Star Upsets Fans," *Winnipeg Free Press*, 5 August 1989, 32.

showed, it is important to analyze what conclusion is meant to be drawn from the inconsistency. If the conclusion is taken to be that the environmental cause is in itself false or wrong, case 6.5 could be evaluated as an *ad hominem* fallacy. But there appears to be no textual evidence to support this interpretation.

In still other cases, the circumstantial *ad hominem* argument is used to question the credibility of someone who has made a charge or allegation, the conclusion being that the charge can be disregarded. Here the *ad hominem* argument is used to rebut a weak presumption.

In an article on the brother of Corazon Aquino, Jose Cojuangco, Jr., president of a key political party in Aquino's ruling coalition, it was reported that he was "feeling heat from the scandalmongers."[10] Defense Minister Juan Ponce Enrile had accused Cojuangco of "playing a behind-the-scenes role in the state-owned firm that runs gambling casinos, once almost a private money machine for Marcos cronies who have since fled the country." However, the *Newsweek* article remarked on the "partisan tinge" of these accusations, calling Enrile a "renegade" minister.

Case 6.6

> Enrile, who remarked recently that "a new oligarchy seems to be emerging," is hardly one to talk; during his own long career as a Marcos ally, he prospered in the banking, coconut and logging industries—and his critics would welcome a close look at his books by the Presidential Commission on Good Government.[11]

This rebuttal of the allegation made by Enrile is a double *ad hominem* attack. First, the suggestion is that there could be a partisan political motivation—Enrile is described as a "renegade" Defense Minister. Second, and perhaps more importantly, the argument is that Enrile is not a credible spokesman to lay charges of corruption, because he himself is under suspicion of the very same charges. The article states that Enrile "is hardly one to talk," for during the Marcos government, he "prospered in the banking, coconut and logging industries," and his critics suspect him of violation of ethical principles of good government. The technique used

10. Spencer Reiss, Melinda Liu, and Richard Vokey, "An Equal-Opportunity Rumor Mill," *Newsweek*, 10 November 1986, 38.
11. Ibid.

here is to combat one rumor or suspicion of scandal by raising another rumor against the source of the first one. The thrust of the rebuttal is to the effect that this person does not have any right to make this kind of accusation because he is on equally weak moral ground himself. It seems to be a tactic of fighting rumor with rumor.

The danger in this kind of *tu quoque* circumstantial *ad hominem* is that the dialogue can degenerate into unsubstantiated gossip or "rumor trading." However, as a method of replying to rumors and allegations based on a source's suspicions rather than hard evidence, the circumstantial *ad hominem* argument that questions the integrity of the source is basically reasonable. A presumption can be fairly attacked by posing a counterpresumption.

Bias *Ad Hominem*

In the third basic type of *ad hominem* argument, a critic questions the personal trustworthiness of an arguer by claiming that this arguer is biased. For example, a critic might claim that an arguer has something to gain by adopting the particular position he has advocated on an issue. This third type of *ad hominem* attack is related to the other two because in this type the critic is typically suggesting that the motives of the person criticized are open to suspicion. This relates to the abusive *ad hominem* argument because there may be a suggestion, for example, that the individual cannot be trusted. But it also relates to the circumstantial *ad hominem* because circumstances may be cited from this person's past that show that he is biased or has something to gain by arguing for one side of the issue in this case.

This third type of *ad hominem* criticism is complicated because it suggests that an arguer has a hidden agenda. The suggestion is that the arguer is not as concerned with truth or with his real convictions as he purports to be; instead, he is secretly negotiating in a kind of interest-based bargaining dialogue. Thus, the suggestion in this third type of *ad hominem* criticism is that the individual criticized has made an unwarranted or unannounced dialectical shift from one context of dialogue to another, while he may pretend to be engaged in a critical discussion. The criticism alleges that he is really engaging in a process of negotiation on behalf of his own interests. The criticism, then, suggests that because this negotiation is unilateral and secret, this person cannot be trusted to engage in reasonable critical discussion in a fair and honest manner.

In some cases the bias *ad hominem* argument is appropriate and effective—a reasonable and useful criticism for the situation. In the following case from Walton 1989 (149), Bob and Wilma are engaged in a critical discussion on the problem of acid rain.

Case 6.7

> Bob and Wilma are discussing the problem of acid rain. Wilma argues that reports on the extent of the problem are greatly exaggerated, that the costs of action are prohibitive, and that therefore no actions to control acid rain are needed. Bob points out that Wilma is on the board of directors of a U.S. coal company and that therefore her argument should not be taken at face value.

Here Bob is using the bias *ad hominem* argument to question whether Wilma has a hidden agenda to support one side because she is connected with a company that has a lot at stake financially with the problem of acid rain. Bob is suggesting that Wilma is not really engaged in a fair and open-minded discussion of the issue, as she purports to be. In reality, Bob is suggesting, Wilma is pushing for the one side because she is biased.

Of course, even if Wilma has this financial connection, it does not follow that she has to be always taking the side of the industrial polluters.

Perhaps she is more fair-minded than that. Even so, Bob's criticism has sting because Wilma does not openly say at the beginning of the discussion that she has this personal affiliation. The allegation of concealment raises critical questions about Wilma's sincerity and raises the possibility that Wilma may be unilaterally (and secretly) engaged in a negotiation or quarreling type of dialogue.

Of course, much depends on the original context of dialogue. If it had been an impartial inquiry into the problem of acid rain, Bob's *ad hominem* criticism certainly would be effective and appropriate. But even in a critical discussion, where Wilma is taking a partisan point of view for one side, Bob's revelation of personal interest does have force as an effective criticism. Bob is not saying that Wilma's arguments are totally wrong, but only that her personal circumstances raise questions about her sincerity in looking at the evidence on both sides.

The *ad hominem* criticism is best perceived as a weak refutation, a rebuttal that is not an absolute rejection of the arguer's conclusion but only a challenge to the arguer's objectivity in supporting that conclusion. In case 6.7 above, Bob's circumstantial challenge of Wilma's argument is a weak but relevant criticism. Bob is making a relevant point if his conclusion is that Wilma's argument should not be taken at face value. But because this argument against the person is a weak criticism, Bob would be committing a fallacy if he were to conclude that Wilma's claim must be false, or that her argument must be wholly rejected for the reason he gives. For Wilma's argument can be a good one even if she is biased.

In evaluating case 6.7, much also depends on just how Wilma has distanced herself from, or identified herself, with the interests of the coal companies in her arguments. And much depends too on how Wilma responds to the criticism. An evaluation requires a careful analysis of Wilma's commitments, as expressed by her arguments. In a case like this, Bob's personal attack may be so effective that it will be difficult for Wilma to contend with it. Unless his allegation is not based on good evidence and Wilma can show that the evidence is weak or nonexistent, Bob's personal attack is sure to have a poisoning-the-well effect on the audience.

In effect, Bob's personal attack suggests that Wilma's arguments that the acid rain problem has been greatly exaggerated and that therefore no action is needed cannot be trusted, because she is a biased advocate for her side of the argument. In other words, Bob is saying that Wilma is not going to be affected by reasonable argument or by objective evaluation of the evidence on the issue. Bob claims that Wilma has so much to gain personally by pushing her side of the argument that she is not a serious or open-minded arguer.

The effect of this charge may be to close off the possibility of continuation of reasonable dialogue. If Wilma cannot be trusted, then her reasons and arguments may be hypocritical and, as such, are not worth paying further attention to. In effect, Bob's argument disqualifies Wilma as a participant in reasoned dialogue—he is suggesting, "There is no point in listening to her any further." The danger with the personal attack argument is that it can seal off reasonable dialogue altogether.

Personal and emotional arguments can be in themselves logically reasonable and appropriate, but even so, they may have a persuasive impact out of proportion to what logic should allow. Consequently, the effective use of an *ad hominem* personal appeal is often a good sign that reasonable dialogue is about to deteriorate into a quarrel or an adversarial debate. Once the poisoning-the-well effect takes hold, it may become impossible to get the dialogue back on track.

Poisoning the Well

The bias *ad hominem* argument is often identified in the textbooks with "poisoning the well." In this type of criticism an arguer is accused of being hopelessly biased and therefore not to be trusted as capable of cooperative, sincere, and reliable participation in a discussion. The strategy of "poisoning the well" is to seal off an argument by excluding the other party.

According to Engel (1976, 109), the term "poisoning the well" was introduced into logical terminology through a dispute in the nineteenth century between the novelist Charles Kingsley and the British churchman John Henry Newman. During the course of one of their arguments, Kingsley suggested that Newman did not place the highest value on truth because he was a Roman Catholic priest. Newman's counter to this was that the accusation made it impossible for him, or for any other Catholic, to take part in the argument with any credibility. In effect, by making this criticism, which poisoned the well, Kingsley had automatically ruled out as hopelessly biased any argument that Newman might bring forward. He had therefore discredited in advance any of Newman's arguments, no matter how reasonable they might be.

Using the label "poisoning the well" to cover the third basic type of *ad hominem* is, however, to some degree, a misnomer and oversimplification. It would be more accurate to call the third basic type of *ad hominem* argument the bias *ad hominem* argument. It is only when the bias *ad hominem* argument is used in a stronger form that it is fair to call it a case of poisoning the well. For it is one thing to criticize somebody's argument by saying he or she is biased, but it is a stronger form of criticism to say that this person is so hopelessly biased as not to be trusted at all to cooperatively take part in the argument or tell the truth. The poisoning-the-well *ad hominem* argument is actually an extension of the bias *ad hominem* argument combined with the additional use of the abusive or personal *ad hominem* argument.

In poisoning the well, the thrust of the argument is that this person is not only biased; he is so determined to stick to this bias that he can never be trusted to argue in an honest manner, can never argue without secretly basing his arguments on this bias. Thus, the poisoning-the-well *ad hominem* argument is a very strong form of criticism because it alleges that the other participant really isn't qualified or reliable to take part in an honest critical discussion in a cooperative manner. The poisoning-the-well argument is such a strong form of criticism because it automatically discredits everything that an arguer might say in the future course of the argument.

There are several important relationships between the different types of *ad hominem* arguments. The poisoning-the-well variant is typically an extension of the bias *ad hominem*. But it can also grow out of the personal *ad hominem* or be extended from the bias variant with its aid. The poisoning-the-well variant deepens the eristic aspect of a dialogue. After all, if one's opponent is a "lying devil," there is no use at all trying to have a critical discussion that depends on presumptions of honesty and sincerity in communicating together in dialogue. All one can do with a lying devil is to use clever tactics to get the best of him in eristic dialogue. Since his mind is made up on the issue, there is no point in trying to convince him rationally. Any slide toward the poisoning-the-well *ad hominem* forces the abandonment of a critical discussion, replacing it with a quarrel.

One problem with *ad hominem* argumentation generally is that once a certain level of derogatory language has been used to classify one's opponents in the argument, a kind of poisoning the well takes effect. For example, suppose two different groups, the Articans and the Borians, are disputing over some claim, like a territorial border, and the Articans start referring to the Borians as "lying devils who never tell the truth." Once the term gains currency, the Borians are bound to find some equally prejudicial term to refer to the Articans. The use of this terminology will make it impossible for the Articans and Borians to have any genuine critical dialogue or negotiation that is not already, to some extent, a quarrel. The *ad hominem* argumentation is built right into the language of the conversation.

This *ad hominem* problem stands out especially when one side in a dialogue unilaterally launches into personally loaded language implying the guilt or blameworthiness of the other side, and the other side refuses to join in this rhetoric. In the following case, a professor who had been turned down for tenure in the English department of a university took his grievance to the streets in a way that was felt by his colleagues to have been a "failure to observe the niceties of academic discourse."

Case 6.8

He told a crowd of 250 student supporters the full-time professors were: "Europhilic elitists and white supremacists," and labelled them Ivy League Goebbels who could "flaunt, dismiss, intimidate and defraud the popular will." "We must unmask these powerful Klansmen, these enemies of academic freedom, people's democracy and Pan-American culture. They must not be allowed to prevail. Their intellectual presence makes a stink across the campus like the corpses of rotting Nazis," he said.[12]

The departure from standards of academic discourse exemplified by this speech caused this professor loss of support. The chairman of the English department was quoted as saying: "Once you start labelling people Nazis and Klansmen basically you're saying there can no longer be conversation or dialogue."[13] In this case, one side unilaterally went "to the streets" and engaged in an eristic dialogue that was not perceived as an appropriate language for a tenure dispute or academic discussion by the other side.

When one side launches into this type of *ad hominem* rhetoric without the consent, agreement, or equal participation of the other side, the shift is unilateral. For this reason it can be classified as an illicit shift from one type of dialogue to another.

However, when both sides to a dispute launch into *ad hominem* rhetoric of this sort, the problem can be even more difficult to resolve, and it is less obvious that a fallacy or illicit shift is occurring.

Tu Quoque Arguments

In many textbooks the *tu quoque* argument is also classified as a distinct type of *ad hominem* argumentation. For example, van Eemeren and Grootendorst (1984, 190) classify the *tu quoque* as a separate category of *ad hominem* argumentation, identifying it with what is classified above as the

---

12. "Untenured Professor in Nazi Jibe," *The Times Higher Education Supplement*, 20 April 1990, 11.

13. Ibid.

circumstantial *ad hominem* argument. *Tu quoque*, meaning "you too," refers to argumentation where one party has advanced an argument to which the other party replies: "You can't fairly criticize me on that basis, for you are just as bad. You are doing the same thing yourself."

This rejoinder can be effective in argumentation. It is often associated with *ad hominem* arguments where an *ad hominem* is used to reply to a previous *ad hominem*. However, the *tu quoque* can be used with many different kinds of criticisms and is not restricted to *ad hominem* argumentation. It is therefore not a type or subspecies of *ad hominem* argument, even though it is worth remarking in textbook treatments that the *tu quoque* is often associated with *ad hominem* argumentation. For example, in the abusive *ad hominem* attack, when one party attacks another party's character for veracity or morality, the person attacked may quite often turn around and say, "Well, you're just as bad," or "You're just the same," or "You've done the same thing in the past yourself."

The *tu quoque* is also often used in connection with the circumstantial *ad hominem* argument. In the *tu quoque* form of this argument, the criticism is made, "You criticize me for doing this particular action, but then you yourself do it as well, so you're just as bad." However, when this rejoinder occurs in argumentation, it should not be classified as a separate type of *ad hominem* argument but rather as a use of the circumstantial *ad hominem* argument to reply *tu quoque* to someone else's use of the circumstantial *ad hominem* argument. That is, the *ad hominem* argument generally is an argument that can be turned on its head effectively by someone who is criticized by it. This does not mean, however, that the *tu quoque ad hominem* argument is a special argument in its own right. It just means that the *ad hominem* argument, being based as it is upon a criticism of an individual's personal circumstances, can often be effectively turned on its head by directing the very same argument back against the critic.

In the following case from the Oral Question Period in parliamentary debate (*Canada: House of Commons Debates*, 5 February 1986, 10469), the Honorable Donald J. Johnston of the opposition Liberal party, posed an *ad hominem* question. The subject of the dialogue was interest rate levels. In the previous exchanges, the government Finance Minister, the Honorable Michael Wilson, had argued that the dollar was "up" rather than "down" (10468). During this exchange Mr. Johnston quoted Mr. Wilson having said "that the dollar is undervalued and that it will rebound." During the segment quoted below, Mr. Johnston uses the *ad hominem* technique of holding Mr. Wilson to his expressed commitment on pain of appearing to go against his own previous commitments.

Case 6.9

Request for Assurance from Minister

Hon. Donald J. Johnston (Saint-Henri-Westmount): Will the Minister assure Canadians and Members of the House that there will be no increase in interest rates tomorrow, given his belief that the dollar is undervalued and that it is rebounding?

Hon. Michael Wilson (Minister of Finance): Mr. Speaker, this is a ludicrous question coming from the Hon. Member who was a Minister at the time the previous Government was pushing interest rates up to 20 and 25 per cent per annum. There is no credibility whatsoever in that type of question.

Some Hon. Members: Here, here!

Mr. Wilson (Etobicoke Centre): As has been demonstrated, our policy is to get interest rates down. We are encountering pressures in the exchange markets which overlap into financial markets here in Canada, which has resulted in interest rates going up in Canada. However, as we indicated at this time last year, our policy is to get interest rates down by stabilizing markets. That is what we are doing right now, and we will be following through with a reduction of interest rates in due course.

Mr. Wilson suggests *tu quoque* that Mr. Johnston has no right to accuse anyone of raising interest rates when his own government, when Mr. Johnston was minister, was "pushing interest rates up to 20 and 25 per cent per annum." This is a classic *tu quoque* reply.

Is Mr. Wilson's *tu quoque* reply reasonable? To evaluate this, one needs to look carefully at Mr. Johnston's question. The question asks for Mr. Wilson's assurance "that there will be no increase in interest rates tomorrow," given Mr. Wilson's previous commitment to a "rebounding" dollar. Could Mr. Wilson answer such a question directly? Could he either give such an assurance or say directly that he would not? It would seem implausible that he could, for even if he could arbitrarily fix interest rates on a day-to-day basis, he certainly couldn't make public announcements about it, given the effects this would have, for example, on the stock market. The question is not a reasonable one and the questioner and respondent both know it.

Given the context of the question in the given situation, it is quite reasonable for Mr. Wilson to respond that there is "no credibility" in this type of question and to back up this response by challenging his questioner's sincerity and seriousness using a *tu quoque ad hominem* argument. Following this initial response, Mr. Wilson clarifies the situation by outlining his party's policy on getting interest rates down.

A danger inherent in the use of *tu quoque* argument is that even though an *ad hominem* reply to a prior *ad hominem* attack may be reasonable in itself, a slide toward personal argumentation may be hard to reverse once it gets going. A dialogue that starts out as a critical discussion of an issue may get more and more personal until it degenerates into a personal quarrel.

In such a case there is a dialectical shift from one context of dialogue to another. The personal *ad hominem* attack may be perfectly appropriate to a quarrel, but if the original dialogue was supposed to be a critical discussion, the arguments should be evaluated in relation to the goals and rules of the critical discussion. By this standard, the *ad hominem* argument may be fallacious.

In a critical discussion, both parties must be open to changing their points of view if the other party presents a convincing argument that shows that a previous point of view can be refuted by good evidence. In a quarrel, however, this openness to defeat never really exists, for the quarrel is a purely adversarial type of dialogue exchange where the goal of each party is to defeat the other party at any cost.

The ancient philosophers used the term "eristic" to refer to this purely adversarial type of argumentation dialogue, contrasting it with "dialectic," their name for reasoned critical discussion.

Eristic Dialogue

Eristic dialogue has three essential characteristics: (1) it is purely adversarial—attack and defense are all that really matter, and there is no willingness to observe the rules of critical discussion; (2) there is no goal to seek the truth or to discover knowledge; and (3) there is no openness to be won over or defeated, and both opponents intend to remain adversaries no matter what happens in the course of a dialogue. All three of these characteristics show a sharp contrast between eristic dialogue and persuasion dialogue. In persuasion dialogue, there must be a willingness to be persuaded by the other party and to change one's commitments if sufficiently persuasive arguments are presented by that other party. In persuasion dialogue there should be regard for the truth of the matter being discussed. And especially in the critical discussion type of persuasion dialogue, there must be a following of the proper rules and procedures at each stage of the discussion. In eristic dialogue, by contrast, the aim is to get the best of the other party by any means, foul or fair.

Eristic dialogue often appears, on the surface, to be like a critical discussion in some respects, because it is common in eristic dialogue to make a show of presenting a "logical" argument and to criticize the other party for self-contradiction or other violation of the rules appropriate for critical discussion. But this quite often is all a sham, a way of "blaming" the other party for being "illogical," while one's own side is being dispassionate and logical. It is a tactic for trying to occupy the higher moral ground in the dialogue—often a deceptive and unfair tactic that involves duplicity and pretence.

The quarrel is a subtype of eristic dialogue, but it is a special subtype that is paradigmatic of eristic dialogue. The goal of the quarrel is to "hit out" verbally at the other party and, if possible, to humiliate the other party. It is possible to speak of a physical quarrel, but in connection with argumentation the term *quarrel*, as a type of dialogue, signifies a verbal exchange.

The quarrel is a personal and emotional type of dialogue. It begins with an emotion of truculence or "hurt feelings" on at least one side. This level of emotional response is typically escalated, rising to a breaking point during the course of the quarrel. The aim of arguing in a quarrel is personal attack—each party tries to "counterblame" the other party for guilty actions or character defects. Hence the personal attack argument or *argumentum ad hominem* is typically associated with the quarrel as a type of dialogue and is a key sign of the existence or opening of a quarrel. For this reason, the quarrel is often called "the personal quarrel," but there can also be group quarrels.

The main benefit of the quarrel is a violent release of pent-up emotions, called the *cathartic function* of the quarrel. Another valuable benefit of the quarrel is that it can serve as a substitute for physical fighting, a way of achieving catharsis without the damages associated with physical fighting. So the quarrel is not inherently bad or fallacious in itself as a type of dialogue. Fallacies occur where another type of dialogue, like a critical discussion, shifts into a quarrel.

The debate is often confused with the quarrel or portrayed as a kind of quarrel, but these two types of dialogue are quite distinct. The quarrel is a personal exchange between two parties. The debate is an exchange between two parties where the goal is to win over, or persuade, a third party audience (or judge or referee, or whoever is to decide who has won or lost the debate).

The debate does have a partially eristic character. Its basic principle is survival of the fittest. Each side is allowed to bring forward its strongest arguments and rebuttals in the "bear pit," and the audience is allowed to judge which side has survived the onslaughts of its opponent's attacks and come out as the better argument in the end. The point is to allow an adversarial contest.

Another essential aspect of the debate is containment of the antagonistic techniques by the rules of debate, clearly laid down at the outset and agreed to by the participants. The extent to which a debate exhibits rational argumentation as opposed to quarreling depends on the nature and enforcement of the rules. Often, the rules are boundary conditions that exclude certain explicit types of attacks and moves but are generally quite permissive. For example, in parliamentary debates, time limits are usually tightly enforced and certain types of attacks—like calling a member a liar—are excluded; at the same time, all kinds of questionable tactics and dubious arguments are allowed, by the rules, to be advanced. If they are bad arguments, it is up to the opposing side to point this out. Thus, debates typically do allow for all kinds of antagonistic attacks back and forth, and it is not hard to see why the debate is often perceived as a type of quarrel.

It is also possible to have mixed dialogues that are partly quarrel and partly debate. The kind of argumentation called the "adversary argument" by Flowers, McGuire, and Birnbaum (1982, 275) can be classified as an eristic debate, a type of mixed dialogue that is partly quarrel and partly debate. In this exchange the participants, who intend to remain adversaries, present their arguments for the judgment of an audience, and the aim of each arguer is "to make his side look good while making the opponent's look bad." According to this account, personal attacks play a central role in adversary arguments, whereby each side tries to imply that the other side is "bad" or guilty for having committed some blameworthy acts. The example they give is that of an Arab-Israeli dispute about who started the 1967 war. Each side tries to cite "bad" episodes the other side was responsible for and that led to the war.

In the terms here, this type of dialogue is classified as an eristic debate because the aim is to defeat the other side, and neither side is really open to being persuaded, but the exchange is directed to third-party onlookers and is meant to be for their benefit and to be judged by them. So this type of dialogue has some of the properties of the quarrel and some of the properties of the debate.

The Shift to the Quarrel

The *argumentum ad hominem* is associated with dialectical shifts from one context of dialogue to another and, in particular, with shifts to the quarrel. In some cases it is even possible to have a *cascading effect*, a series of shifts. For example, there could be a cascading from a critical discussion to a debate, to a negotiation, and then to a quarrel.

Some dialectical shifts can be *licit*, meaning that the shift is not bad or a deterioration of the original dialogue. Indeed, in some cases one dialogue can be functionally embedded in another, so that the second one actually contributes positively to the goals of the first. An example would be a critical discussion on the harmfulness of acid rain that is enhanced by the insertion of an expert consultation dialogue where scientific experts in fields related to the problem introduce relevant information. This would be a licit shift.

Other shifts are *illicit*, in the sense that they represent a deterioration, where the goals of the old dialogue are hindered or blocked by the advent of the new dialogue. Usually, a shift to a quarrel from another type of dialogue is suspicious and likely to be illicit.

The shift from the negotiation to the quarrel often takes the form of an escalation where bad feelings on both sides become more and more centered on the personality of the adversary. Eventually, as the quarrel predominates the dialogue, there are extreme emotional outbursts on both sides.

A good example of this dialectical shift occurred in 1989 after Frank

Lorenzo, chief executive officer of Eastern Airlines, tried to cut costs by cutting wages, with the result that the airline workers went out on strike. According to the account given in Newsweek,[14] the unions began to portray Lorenzo as a symbol of ruthlessness and capitalistic greed, even booing at his picture when it appeared on television, calling out, "There's the slimeball!" A union memo urged members to "make Frank Lorenzo *the issue*" and to "personalize the conflict" between Lorenzo, as the "pillager of the American Dream," and the "ordinary working people" of America.[15]

As the struggle between Lorenzo and the union became more bitter and emotional, the personal vilification of Lorenzo as the hated enemy and oppressor became so overwhelming that any possibility of further concessions in negotiation was forgone. According to the Newsweek account, the situation could have been compared to the conflict in Lebanon.

Case 6.10

> With Eastern already swimming in red ink when he bought it, he had no choice but to slash costs. But just as the workers grew obsessed with Lorenzo, he became consumed with the single-minded strategy of beating the unions. The fight became so personal that any possibility for compromise was lost. "This is very much like the situation in Lebanon," says Richard D'Aveni, professor of corporate strategy and organizational design at Dartmouth's business school—"with factions that hate each other so badly nothing will solve the problem."[16]

Eventually, the negotiation degenerated into a quarrel, as each side became "obsessed" and "consumed" with the single-minded goal of attacking and defeating the other side of the struggle.

This case is a classic instance of the *glissement*, or gradual shift, from the negotiation dialogue to the quarrel. It illustrates the danger of the personalization of argument, where the issues originally being discussed fade gradually into unimportance for the participants as the personal attacks become so heightened and intense that they eventually choke off

14. John Schwartz, Erik Calonius, David L. Gonzalez, and Frank Gibney, Jr., "A Boss They Love to Hate," Newsweek, 20 March 1989, 20–24.
15. Ibid., 22.
16. Ibid., 23.

the original dialogue altogether. As emotions are intensified, the urge to hit out personally at the other party becomes all-consuming, and no room is left for the give and take of real negotiation.

It would appear natural to diagnose this case as an instance of the *ad hominem* fallacy. However, van Eemeren and Grootendorst (1984) define a fallacy as a violation of a rule of a critical discussion. By their criterion, case 6.10 would not be classified as an instance of the *ad hominem* fallacy or even as a fallacy at all. The reason is that the dialogue in case 6.10 was supposed to be a negotiation, not a critical discussion.

What is one to say here? Is the argument in case 6.10 an *ad hominem* argument, but not an *ad hominem* fallacy? Could it be classified as some problem or failure of communication other than a fallacy?

It seems very natural to classify case 6.10 as an instance of the *ad hominem* fallacy. It illustrates precisely the danger in the use of *ad hominem* argumentation as a personal attack escalates into a quarrel. True, the original dialogue was a negotiation and not a critical discussion. But the danger of the *ad hominem* in destroying a negotiation by escalating it to a quarrel is a danger well worth warning about and categorizing as a type of *ad hominem* fault or problem.

Should the shift from negotiation to quarrel be included as a fallacy? Or instead, should a case like the one above be called a problem or failure of negotiation that is not an instance of the *ad hominem* fallacy? The proposal put forward here is to classify case 6.10 as an instance of the *ad hominem* fallacy. The acceptance of this proposal implies that it is possible to have a fallacy where there has been a shift from a negotiation dialogue to a quarrel.

The *ad hominem* fallacy occurs when there has been an illicit shift from a prior type of dialogue to a quarrel, because the quarrel is not an efficient way of carrying out the goals of the prior dialogue and even blocks those goals altogether.

Personalization of Argument

The articulation of the personal position of a participant in a critical discussion can be assisted by an elenchtic questioner who probes and challenges the commitments of the respondent. This is the *maieutic function* of dialogue, the use of dialogue to give birth to personal insights that deepen one's understanding of one's own position of an issue (see Walton 1984, 247–54; 1987b, 125–29; 1989b, 296–300). The question functions like a midwife to assist the respondent's delivery of her new idea into the world. It is the respondent who produces the idea that was all the time contained in her, and the questioner's function is to assist in this process of delivery.

To model this maieutic function of dialogue, a Hamblin-type commitment set can be divided into two subsets—a *light side* of overtly expressed commitments known to the participants and a *dark side*, a set of propositions that are definitely commitments of the participants but are not definitely known to the participants. One participant does not, in general, know what the other participant's dark-side (hidden) commitments are (see ibid.). But he can make presumptions about what they are likely to be, given the respondent's answers to questions, and inferences that the questioner can draw from what he does know about the respondent's commitments.

One way of extracting dark-side commitments is by posing *ad hominem* questions that ask a respondent whether what she says now is consistent with her other commitments (darkly) conveyed through her previous actions, affiliations, and other personal circumstances.

In some cases, *ad hominem* arguments can be an asset to a dialogue because bringing personal matters into the dialogue may enable a participant to articulate his own personal position on the issue. Such a personalization of the argument can aid the maieutic function of dialogue, whereby an arguer's deeper position is brought out and clarified by the discussion. After all, arguing from an opponent's concessions, the premises she is committed to, is the central method of a critical discussion or any type of persuasion dialogue.

But there is a danger that such a personal exchange can turn into a quarrel. Once personal matters come into it, emotions are heightened, there is a fear of loss of face, and refraining from escalating a personal attack is often difficult.

A critical discussion is improved if it is personal to some extent. But if it becomes too heavily or obtrusively personal, the quality of the dialogue can deteriorate. The real issue may not get discussed, and fighting may result instead. It is exactly in this case that the *ad hominem* argument moves toward becoming a fallacy.

So using *ad hominem* argumentation requires judgment and restraint. It can be a beneficial type of argumentation up to a point. But beyond that point it can become problematic, obstructive, and even fallacious. One must be aware of the critical limits of this type of argumentation and not get carried away to excesses by the personal element.

Many of the cases of the *ad hominem* argument treated in the logic textbooks come from the discourse of political debate. And, as has been shown in Walton 1989b (chapter 5), the *argumentum ad hominem* is a common and also highly powerful and effective type of argument in political debating. However, as has already been seen in Chapter 3, political debate in a democratic system is a complex, or mixed, type of dialogue involving negotiation, critical discussion, eristic dialogue, action-producing dialogue, and information-seeking dialogue. As noted in the section of Chapter 6 on poisoning the well, the debate generally is a mixed type of dialogue containing elements of the critical discussion and elements of eristic dialogue. For these reasons, it can be quite a subtle and complex problem to evaluate the uses of the *ad hominem* argument in political debates. Some cases appear quite simple and can be straightforwardly evaluated, yet pose subtle and difficult problems that have not yet been resolved.

Case 6.11

A Catholic politician running for a high federal office declared that she supported freedom of choice on the abortion issue, even though, as a Catholic, she personally opposed abortion. She argued that her personal views are not in conflict with her position on public policy. A Catholic bishop criticized this stance as illogical, replying that he did not see how a good Catholic, who should be against the taking of human life, could vote for a politician who supported abortion. She replied that as a Catholic she did not personally support abortion, but that she felt she had no right to impose that view on others, who might have different religious viewpoints. She stated that her political support of freedom of choice concerning reproduction was logically consistent with her personal opposition to abortion because of the separation of church and state. (Walton 1989, 169)

This case poses a problem, for commitment to deeply held moral principles of personal conduct surely must not be completely independent from one's political policies on moral issues, even if these policies are questions of public policy, hammered out in the fires of political compromise. Surely there is a difference between one's personal standards of morality or religious commitment and one's support of policies, legislation, or public guidelines that apply to a whole group of people or even a nation. Tolerance and compromise, as well as freedom of religious belief, mean that one's (moral) personal and public (policy) commitments are never going to be absolutely synchronized or consistent with each other.

On the other hand, too severe a circumstantial inconsistency between these two commitments can raise critical questions that shift a burden of proof against a politician to rebut the presumption of "not practicing what you preach." In this case, the bishop's question seemed appropriate—how could this politician, as a good Catholic, take a pro-life position personally while taking a pro-choice position politically? Here the circumstantial *ad hominem* attack, as a form of presumptive critical questioning of an apparently ambivalent position, does seem like a reasonable argument in the context of the political debate. It would seem right that this politician owes some sort of explanation to voters to retain personal credibility.

In discussing this problem Cuomo (1984) emphasized the religious pluralism of American democracy, according to the principles of a constitution that guarantees freedom of religious opinion. Even one opposed to abortion on religious grounds consistently maintains that legislating against abortion is not the best way to act in accord with one's personal commitment to oppose abortion. The question of putative circumstantial inconsistency then reduces to the question of what actions this politician has taken to support the pro-choice movement. Supporting this movement abstractly is one thing. However, joining an action group to establish comprehensive abortion services for women across the country is quite another. Further information would better define the presumptive inconsistency.

What is indicated by this case is that the *ad hominem* attack is best treated as a defeasible criticism based on a presumption of circumstantial inconsistency or personal insincerity, a presumption that can be rebutted by a convincing explanation or clarification of the arguer's position. Such a rebuttal can shift the burden of proof against the weight of presumption raised by the critic. In this case, the politician could conceivably give a good explanation of the apparent inconsistency. But to retain credibility, an explanation is "owed."

Here too the circumstantial *ad hominem* is systematically linked to the personal *ad hominem* argument, the one naturally leading to the other.

Once the allegation of circumstantial inconsistency raises the need to clarify the arguer's deeper personal commitments on an issue, not only is the argument personalized, but questions of the arguer's honesty and sincerity are also raised. On the basis of an alleged circumstantial inconsistency in a person's position, for example, a critic may suggest that this person is a hypocrite, which then raises questions of personal integrity.

A reasonable use of the *argumentum ad hominem* depends on a presumption that the proponent is accurately representing his opponent's position and fairly interpreting her commitments as expressions of that position. But it is easy for a proponent of an *ad hominem* attack to get carried away and to begin to exaggerate and distort the opponent's position in order to make his attack appear stronger and more effective in refuting the opponent's case. Thus the *ad hominem* arguer can easily begin to distort his opponent's position, engaging in "straw man" tactics.

For example, an *ad hominem* attacker may begin to suggest that his opponent is an "enthusiastic zealot," a fanatic who has a closed mind and has been carried away by her emotional involvement in her own cause. Going even further, the *ad hominem* argument may begin to suggest that the opponent is "mentally ill" and has temporarily lost her grip on reality. In such a case, the bias attack has gone a step too far and has reached the stage of a poisoning-the-well argument. The accusation is now that the opponent is incapable of engaging in proper open-minded or rational argument at all and must be "written off" as a serious arguer altogether.

It is at this latter or poisoning-the-well stage of the bias *ad hominem* argument that the shutting down of dialogue occurs. Moreover, such an obstruction of dialogue is often compounded by quarreling when the other side responds in kind with *tu quoque* replies.

A type of argument that can be reasonable in some cases, fallacious in others, the *argumentum ad hominem* is best defined generally as the direct or indirect attack on a person's character for veracity or sincerity in dialogue by citing pragmatic inconsistency or bias of that person in order to raise critical questions about that person's argument. It is a consequence of this definition that an argument that is an attack on a person without being an attack on a specific argument attributed to that person is not a genuine *ad hominem* argument.

A case in point is case 6.2. Was this argument just an attack on Jim Wright's character, or was it an argument to refute some contention of Mr. Wright's? To the extent that there is no evidence that the latter requirement is met in case 6.2, the argument cannot be said to be a genuine *argumentum ad hominem*. This way of defining the *argumentum ad hominem* is bound to be somewhat controversial because in some cases the argument presumed to be refuted is nonexplicit and needs to be extracted from the discourse by reconstruction. However, the definition above is part of the general dialectical definition of the *argumentum ad hominem* given in Chapter 8.

7

BORDERLINE CASES

Although an analysis of each of the arguments involved in the four emotional fallacies has been given, much work remains to be done to show how these analyses apply to the idiosyncrasies of particular cases. Each real case is always unique in certain respects. In analyzing any given case, one needs to make presumptions about the specifics of the argumentation, based on the available evidence from the text of discourse.

In the first half of the chapter, two cases of the *argumentum ad baculum* are evaluated using the analysis from Chapter 5. Even though the evaluation made of these cases has to be conditional, still it can be seen how the analysis from Chapter 5 can be helpful. Both cases are borderline cases, in the sense that they involve an argumentation scheme other than the one for the *argumentum ad baculum*.

The second half of the chapter carries the analyses further by examining additional borderline cases where one fallacy is combined with another. For example, *ad baculum* combined with *ad populum* in the same argument is called *bacpop argumentation*. Another case uses a slippery slope argument as a kind of *ad baculum* appeal. No systematic attempt is made to consider all possible combinations. But these combined types of argumentation are useful in revealing more about how the individual fallacies work and in charting the borderlines of each of the individual fallacies involved. They also give some holistic insights into sharing of common features by the group of four.

The Drug Test Case

It is interesting to ask whether an *ad baculum* fallacy has been committed in the following case, reported in *Newsweek*, 20 July 1990 ("A Drug Test," 3).

Case 7.1

> Government officials in Colombia view the drug trial of Washington Mayor Marion Barry as a key test of America's commitment to slowing down demand for cocaine. The case is front-page news in Bogotá, and, says an adviser to Colombian President-elect César Gaviria, "we are watching it very carefully." Colombia has long maintained that the United States should focus its anti-drug efforts on reducing consumption rather than trying to stamp out production in the Andes. There is little doubt of Barry's guilt in Colombia, says the adviser, and he warns that relations between the two countries could sour if the mayor is acquitted or given a light sentence. If that happens, he adds, it will be hard to convince Colombians that their war against the drug cartels is worth the price of continued violence and bloodshed.

One important type of dialogue involved in this case is the criminal trial. Should this external pressure be allowed to bear on the trial? Diplomatic negotiations are also involved. Hence there is a shift from one type of dialogue to another. It also needs to be determined whether a threat is made, given that the speech act is, at least overtly, in the form of a warning.

There is a core of legitimacy to Gaviria's argument, seen from the Colombian point of view. Colombians, according to Gaviria, are suffering negative effects—"continued violence and bloodshed"—from the U.S. efforts to stamp out drug production in the Andes. Therefore, according to Gaviria, Colombians naturally question how sincere the U.S. is in its declared "war against drugs," given that it so often seems that they are not serious in their efforts to reduce internal consumption. The Colombians, according to Gaviria, see the handling of the Barry trial as a case in point, "a key test of America's commitment to slowing down demand for cocaine." If Barry is acquitted or given a light sentence, it raises questions, in Colombian eyes, of the sincerity of the U.S. when they claim that they are committed to reducing cocaine usage in their country.

All the elements of a reasonable use of the circumstantial *ad hominem* argument are present here. The U.S. claims, in their rhetoric, that they are committed to reducing drug use in their country. But if so, then they should make serious efforts to reduce consumption by enforcing their laws against drugs like cocaine. But are they really doing this—"practicing what they preach"? If they "go soft" on the Barry case, making an exception of it, critical questions are raised on this score. In this *ad hominem* argument, the question is raised whether the U.S. agencies that (at least verbally) take a strong stand against illegal drug use are truly, sincerely committed to what they say, as shown by their actual practices. As a raising of critical questions about an arguer's commitment in dialogue, the *ad hominem* argument is here being used in a legitimate and reasonable way, focusing on the handling of the Barry trial as a case in point.

On the other hand, Gaviria's argument veers towards fallacy at the later stages, when the *ad baculum* tactic is introduced. It is an *ad baculum* argument because there is a threat involved, and the threat is used as part of Gaviria's argument to press forward with his case.

The threat is indirect—an indirect speech act. Gaviria warns that relations between the U.S. and Colombia could sour if Barry is given a light sentence or acquitted. This is overtly a warning, but covertly it is a threat.

The essential condition for the making of a threat is present. The speaker, Gaviria, is the President-elect of Colombia. So there is reason to believe that he is in a position to bring about the bad outcome of souring relations with the U.S. Of course, he does not say (directly) that he will bring about this outcome. But he does say that "it will be hard to convince Colombians" that continuing the war against drugs is "worth the price."

This speech act is similar to the case where the union leader says, in union-management negotiations, "If you do that, the workers will go out on the line!" By suggesting that the workers will take things into their own hands, the union leader is disclaiming that he makes a threat. He can defend himself by replying that it is only a prediction or warning. But in fact this utterance has the force of a threat and leaves open the possibility that the union leader is really saying that he will see to it that the workers go out on strike unless the management negotiators agree not to perform the action in question.

In both cases a kind of optionality or equivocation is involved. The hearer is left with the option of interpreting the speech act as a warning or a threat. Which way should it be taken? Well, of course, in these kinds of cases, it is prudent to take it as being, at least potentially, a threat.

Does the warning in this case meet the three requirements for the speech act of making a threat? The sincerity condition is met, because the cooperation of the Colombians is needed for the U.S. to fight the importation of drugs, and a "souring of relations" between the U.S. and Colombia would undermine that cooperation. The U.S. would want to avoid such a "souring of relations" if possible and would take steps to do so if necessary.

The preparatory condition is met because, as noted above, the President-elect of Colombia is in a position of power. U.S. officials have reasons to believe that he could be in a strong position to bring about such a "souring of relations" or, alternatively, to see to it that such a development was prevented.

But is the essential condition met? Is Gaviria making a commitment to see to it that such a "souring of relations" will occur unless the U.S. officials comply by seeing to it that Barry gets a tough sentence? This is where the indirect speech act comes in. Gaviria is not overtly saying this, but of course what he says can and should be taken by U.S. officials as a covert message to this effect.

Pinning Down a Case

What is the context of dialogue in the Barry case? At least initially, it seems that the dialogue is a diplomatic negotiation between Colombian and U.S. government officials. In this context it could be that an indirect threat in the form of an overt warning speech act is not out of place. At least, there is not such a strong presumption that this type of speech act is an *ad baculum* fallacy as there would be if it were to occur in a critical discussion. In negotiations, covert (indirect) threats are quite commonly made as part of the argumentation, and they need not be so contrary to the aims of the dialogue that they can be automatically or quickly presumed to be fallacious.

However, there is also a secondary context of dialogue involved: the criminal trial. The case shifts between one type of dialogue and another. Gaviria is using the indirect threat in the negotiation dialogue to bring pressure to bear against those who are conducting the trial of Marion Barry. Should this kind of external pressure be allowed to bear on the conduct of the trial? From the point of view of the trial itself, as a dialogue with its own rules and procedures, this kind of external, political pressure could rightly be perceived as bias, an interference that could block a fair trial. From this point of view, the shift from the threat in the negotiation dialogue to the secondary dialogue of the criminal trial, through the use of the *ad baculum* argument to connect the two dialogues, can rightly be presumed to constitute the committing of a fallacy. By resorting to the tactic of an indirect threat, Mr. Gaviria's argument goes strongly against the goals of the criminal trial, where the jury (or judge) needs to decide on the question of guilt or innocence, independent of external political pressures and diplomatic negotiations.

In the Barry case it is worthwhile to note that the Colombians are not cognizant of all the facts and are in no real position to judge whether he is innocent or guilty. Nor are they in any real position to decide what his sentence should be. Their evident concern, however, is not with these questions but with the public image of his case, the consequences of how his case is decided with respect to public perceptions. Their argument suggests that Mr. Barry should receive a severe punishment in order to send out a message to Colombians and others that the U.S. is serious in fighting the problem of drug consumption.

According to Gaviria, there is "little doubt of Barry's guilt in Colombia," but on what basis are they making this presumption? Under the U.S. system of justice, an accused person is presumed innocent until a trial proves otherwise. It is not up to Colombians to decide on Barry's guilt or punishment, and putting the U.S. legal system under pressure to decide these questions one way or the other is not a legitimate method of argumentation to arrive at or influence a resolution of the issue.

Case 7.1 illustrates very well the practical difficulty inherent in pinning down real cases of the *ad baculum* argument. It seems impossible to prove conclusively that Gaviria is making a threat. And indeed that is the clever feature of the indirect *ad baculum*. It is an indirect speech act that is overtly a warning and only covertly a threat. By the nature of this covert tactic, the speaker always has an avenue of plausible deniability open: "Why, I was not making a threat. I was only offering a warning to you about what might happen" (argumentation from consequences). Thus, in this type of case, there are inherent difficulties in proving that the speaker has committed an *ad baculum* fallacy.

This case illustrates how the determination of an *ad baculum* fallacy is contingent on the interpretation of a speech act in a given context of dialogue. Such a judgment is tied to one's ability as a participant in natural language dialogue to recognize what amounts to a threat. In practical terms, the ability to recognize a threat and interpret it correctly is a skill that is useful and (in some cases) necessary for survival. As the Barry case shows, however, it is something else again to be able to prove conclusively to a third party that one participant in dialogue has made a threat against another.

In such cases, then, it is wise to recognize a conclusion that defeasibly presumes that a threat has been made, relative to the evidence derived from the interpretation of a text of discourse in the given case.

The Mouse Trap Case

The following case was an advertisement for d-CON mouse bait in *Newsweek* (23 October 1989, 81). The very large headline read, "A Trap Can Catch More Than a Mouse," and the ad also featured a picture of a Lyme-disease tick, magnified seven times, and a picture of a mouse trap. The main body of the ad is quoted *verbatim* below.

Case 7.2

> Mice can carry ticks that carry disease.
> Even dead mice.

When you pick up a trap, wherever it's been set, you may be picking up more than a dead mouse. There may also be disease-carrying ticks. And fleas. And mites. These parasites can carry serious diseases such as Rocky Mountain Spotted Fever, Colorado Tick Fever, Tularemia, Typhus.

Recently, you may have read about Lyme-disease ticks carried by the white-footed field mouse. People bitten by the Lyme-disease tick can suffer temporary paralysis of the facial nerves, pain in the joints, and even severe neurological problems similar to multiple sclerosis.

So if you're a homeowner, be careful in your garage or in any barns or outbuildings you may have. While the mouse that carries Lyme-disease ticks has not been shown to enter homes, the ordinary mice found indoors can carry ticks that also may be as threatening.

Why use traps?

If a trap happens to catch a mouse, the only way for you to dispose of it is to go near it. And that may increase your risk of being exposed to ticks, fleas or mites that can remain alive—even on a dead mouse! Why take chances?

With d-CON bait products, you never go near a mouse.

If you have mice in your home, you'll feel more secure using d-CON bait products instead of traps.

All you do is set out the bait. Mice eat it, then leave and go off to die, without you ever having to touch them. Without you ever having to go near them. Or the ticks they can carry.

The ad concludes by giving names of d-CON products and a mailing address where one can write for more information.

Is there a threat made in this case? The answer is "no," because the essential condition is not met. The test question is the following. Did the advertisers indicate their willingness or intention to bring about the bad outcome (Lyme-disease) for the reader? Clearly the answer is "no," and therefore no threat is made in the advertisement.

It follows then that if the argument in case 7.2 is an *ad baculum* argument, it is a scare tactic (appeal to fear) type of *ad baculum*. And it does seem to qualify for this category. When the ad goes on to cite in graphic detail the effects of Lyme-disease, "temporary paralysis of the facial nerves, pain in the joints, and even severe neurological problems similar to multiple sclerosis," it does appeal to fear. There is no disputing that this list of symptoms may be quite accurate. But the question here is how it is being used in the argument of the advertisement. It does appeal to fears that most readers would likely have and perhaps be influenced by.

One interesting feature of case 7.2 is the use of an *argumentum ad ignorantiam* to support the *ad baculum* argument by conveying a conditional warning: "While the mouse that carries Lyme-disease ticks has not been shown to enter homes, the ordinary mice found indoors can carry ticks that also may be as threatening." Here the argument based on a supposition of ignorance—"While the mouse that carries Lyme-disease ticks has not been shown to enter homes . . ."—is used to open way for a presumptive may-conclusion that takes the form of a warning: that the ordinary mice found indoors can be as threatening as those found outdoors.

Here the *ad ignorantiam* argument functions to shift presumption, based on the practical factor of safety. The argument begins by conceding that there is no proof that the Lyme-disease carrying mice enter homes; even so, on grounds of safety and prudence, one should be cautious about this open possibility. And even worse, those indoor mice could contain equally threatening ticks. Hence the practical advice in the next paragraph: you could be exposed to ticks, fleas, or mites on a dead mouse—so why take chances? This is a rhetorical question that indirectly states a prudential conclusion—if you are careful and wise, you won't take chances. Therefore, you'll buy d-CON bait!

The chain of argument throughout the advertisement is a sequence of practical reasoning directed toward the conclusion that you ought to buy d-CON bait products instead of a mousetrap if you have mouse problems in your house. As is typical of practical reasoning used to influence deliberation by giving advice, the conclusion is an imperative that directs the hearer to a particular course of action as prudentially wise in a given situation. Here, the practical advice takes the form of a warning against a danger to the hearer's health and well-being.

The context of dialogue in this case is that of a commercial advertisement, where the proponent of the argument (the speaker) is trying to get the respondent(s), the reader(s), or hearer(s) to buy a product. It is a kind of persuasion dialogue, and the manufacturer of d-CON products pays to put the advertisement in the magazine to sell its products for profit. The proponent has a bias, in other words, and the purpose of the message is advocacy.

However, at the same time, the message takes the form of advice-giving dialogue using practical reasoning to warn the reader of possible negative consequences, giving the reader (supposedly) practical advice on how to avoid these consequences. It is a mixed dialogue, with one level of dialogue superimposed over the other. Nothing is inherently fallacious *per se*, however, with this mixture of two types of dialogue.

So far, then, although case 7.2 involves a mixture of *ad baculum* and *ad ignorantiam* argumentation, there are no decisive and good grounds for concluding that the argument is fallacious.

Evaluating the Appeal to Fear

This case is a persuasion dialogue, but it is also a commercial advertisement. One might criticize the argument for appealing to fear instead of providing relevant facts about the cost of the product and its chemical properties, including its potential danger to humans and so forth. But commercial advertisements are not normally expected to provide this kind of information, at least in specific detail. To the criticism that he or she is committing a fallacy of irrelevance the advertiser can reply by arguing that the purpose of a commercial advertisement is to draw the attention of consumers to the product. In such a message the point is not to present all the facts but to announce the availability of the product, so that the consumer can look at the facts, decide whether to look into it, and compare it with the other products on the market that are designed for the same use. A commercial should present some information, but there are limits on how far it has to, or should go, in this direction.

In this case the appeal to fear pervades the argument. For example, the picture of the Lyme-disease tick in the advertisement has been "magnified seven times," according to a caption under the picture, making it appear to be a menacing creature. Moreover, some implausible claims about the effectiveness of the product are made, and little or no convincing evidence is given to support them. For example, it is claimed that a mouse killed with d-CON will find a convenient place to die, outside your house, where you won't have to get rid of it. This claim is implausible, and no evidence is given to support it. So the argumentation in the advertisement could be criticized for appealing to fear as a filler or cover-up for the failure to give any real information on how the product works or to give convincing arguments that the product would in fact work as well as it is claimed in the ad. The tactic is to bolster up a weak and implausible argument by appealing to fear.

Should the use of this tactic be classified as a fallacy in this case? What should be criticized is the use of appeal to fear to sell the product without giving useful information on how or why the product is supposed to work. By not giving evidence of this sort, the argumentation in the advertisement does not assist the reader to ask the appropriate critical questions that would help her or him to arrive at a practical decision when comparing this product to another.

If the scare tactics were used to emphasize or highlight an argument that convincingly argued for the usefulness of the product, the *ad baculum* appeal would not be so suspicious. But in this case it seems that the appeal to fear is used more as a tactic to avoid or head off possible rebuttals and critical questions. This provides some grounds for treating it as a fallacious use of the *argumentum ad baculum*.

On the other hand, the ad does give an address to which one can write for a brochure that gives "more information about how to get rid of mice." In the format of a commercial advertisement it is legitimate to make claims about the effectiveness of your product and about the superiority of your product over that of your competitors. And since it is clear to everyone that you are selling your product, you are not expected to represent an impartial, nonbiased argumentation. It is clear that you will be emphasizing the strong points of the product to get the consumer to see it in a good light. A commercial, therefore, is not the place for a scientific study that must objectively present the evidence on all sides. These factors provide grounds for defending this ad against the charge that it uses scare tactics, that it is a fallacious use of the *argumentum ad baculum* of the appeal to fear type.

Hence, in case 7.2, despite the initial tendency to leap to the conclusion that an *ad baculum* fallacy has been committed, because there is an appeal to fear used in the ad, it is better to judge that on balance the appeal to fear is insufficient grounds for such a conclusion. The ad in case 7.2 is appealing to the reader's fear of a disease, but there are grounds for this fear. Whether the ad is trying to exploit this fear by exaggerating it is hard to say. But this judgment need not be made. It is enough to conclude, *ad ignorantiam*, that since the reader cannot know that the appeal to fear is exaggerated, and since there are grounds for thinking it may not be, the reader should give the ad the benefit of the doubt, operating on the presumption that the argument in the ad is not fallacious.

It is best to evaluate case 7.2 by saying that the ad shows a bias in favor of its product and tries to appeal to the reader's bias toward his or her own safety. But, in themselves, neither of these biases constitutes sufficient evidence of a fallacy having been committed.

Judging Individual Cases

In evaluating case 7.2 it may be well to compare it with some reported cases where appeal to fear has been used as a high-pressure sales tactic. According to a report in *Consumer Reports* ("The Selling of Fear: Emergency Help for the Elderly," January 1991, 5), some companies are using appeal to fear to sell emergency-response devices to elderly people. These devices are used to summon help quickly in the event of fall, sudden attack, or other life-threatening emergency.

> In one ad for the *Lifecall* system, a gray-haired woman falls, presses the panic button worn around her neck, and shouts, "I've fallen and I can't get up!" Paramedics and doctors respond on the double.

Although the ads for these systems appeal to fear, the fear, and the appeal to it in order to advertise the product, could be based on reasonable grounds. However, in other cases, it appears that the appeal to fear is being exploited in a questionable way.

> Some companies are using high-pressure tactics to unload merchandise. In one case, recounted by an investigator in the San Francisco district attorney's office, the salesperson reportedly used grisly details about a fictitious crime to sell some systems.

It was reported as well that in some of these cases elderly people had been pressured into buying expensive, but useless, equipment.

What needs to be remembered in judging cases of this sort is that there is going to be differing severity. The appeal to fear should not be judged inherently fallacious. An appeal to fear is but one sign, often the first sign, that an *ad baculum* fallacy may be a fault of the argument. But what needs to be evaluated is how the appeal to fear is being used as a tactic of argumentation in the given case. Only if it is being used to cut off legitimate critical questioning or otherwise thwart the goals of dialogue the participants are supposed to be engaged in, can one conclude that the argument in the given case is fallacious.

Judging commercial advertisements in the media, or other sales pitches, the buyer needs to recognize that the seller is biased and should not be relied upon to present both sides of the issue objectively and critically. While selling has elements of the critical discussion, it also has elements of negotiation and of action-oriented dialogue, with the goal of selling the product. This is not to say that fallacies can't occur in such a context of dialogue. But it is to say that it would be naive to routinely classify appeals to fear as fallacious in this context without paying careful attention to the critical details of how the appeal to fear is used in relation to the goals of the dialogue the participants are supposed to be engaged in.

With many cases of this sort, it is valuable to recognize that a threat or appeal to fear is a significant source of potential critical mischief, an opportunity for logic to start going wrong. But because of insufficient evidence on how the appeal is exactly being used in the context of dialogue in the given case, one must stop somewhere short of concluding that a fallacy has been committed. In cases like this, one should say that there is danger an *ad baculum* (or some other type of fallacy) could be committed if the participants are not very careful. This claim should be treated, however, not as an indictment of fallacy, but as a warning that there exists the danger or possibility of a fallacy being committed in the future sequence of dialogue.

Unfortunately, the traditional tendency on the part of informal logic textbooks has been to jump the gun in these kinds of cases, declaring that an *ad baculum* fallacy has been committed as soon as any threat or appeal to fear has been identified. This simplistic view of the fallacies needs to be overcome.

Guilt by Association and Poisoning the Well

An effective poisoning-the-well *ad hominem* tactic is to set the issue in the framework of an existing or potential group quarrel, at the same time suggesting that the other side is inevitably biased because of membership in the opposed group. The technique in this type of argumentation is to imply that those opposed to one's own point of view are so heavily and hopelessly biased by their own prejudice or fanatical attitude that their opinions can be disregarded altogether, no matter what they might say. This *ad hominem* move, if successful, cuts off all possibility of further critical discussion of the issue in question. There is no room left for looking at the arguments of the other side on their merits, or on good evidence, because the presumption is that the other side must be quarreling. Therefore, they are not open-minded and honest participants in a critical discussion. They are dogmatic, eristic arguers whose arguments can be rejected in advance as biased and one-sided.

A case of this poisoning-the-well *ad hominem* argumentation occurred during a debate on the issue of abortion in the Canadian House of Commons (*Canada: House of Commons Debates*, 30 November 1979, 1920) when the speaker interjected as follows.

Case 7.3

> I wish it were possible for men to get really emotionally involved in this question. It is really impossible for the man, for whom it is impossible to be in this situation, to really see it from the woman's point of view. That is why I am concerned that there are not more women in this House available to speak about this from the woman's point of view.

The implication made by this *ad hominem* argument is that men cannot see the issue of abortion from the woman's point of view; therefore whatever stance they take, it is bound to be hopelessly biased. The suggested conclusions are that whatever men say on the issue can be disregarded as irrelevant and that the opposition to the view favored by the speaker can be dismissed as bias.

The problem with this type of *ad hominem* attack is its divisiveness, the pressure it exerts to set a group quarrel in place. Once the women conclude that it is impossible for the men to see the issue from a woman's point of view, it is of course natural for the men to presume (*tu quoque*) that it is impossible for the women to see the issue from anything other than a woman's point of view, to conclude that the women have shut off any other point of view and are hopelessly biased. Once the initial *ad hominem* arrow has flown it is easy for both sides to settle into the defensive fortifications of quarreling. But where there is no longer an openness to consider the point of view of the other side as reasonable, there is no longer any real chance for a critical discussion of the issue to take place.

This *ad hominem* tactic poisons the well by citing or imputing membership or affiliation in a group opposed to one's point of view and in so doing tries to incite a quarrel.

This fallacy would appear to be a special type of what has been called the *fallacy of guilt by association*, which "occurs when we dismiss a person and their views or arguments on the grounds of their alleged association with some group or cause" (Groarke and Tindale 1986, 310). The fallacy of *poisoning the well by inciting a group quarrel* outlined above is a special case of the fallacy of guilt by association because in the former fallacy, not only are the person's arguments dismissed on the grounds of group association, but it is also alleged that because of this association, the person's arguments must always be hopelessly biased and can never be trusted as sincere or reliable. What is suggested in the latter fallacy is that the person in question is not really open to rational persuasion.

The guilt-by-association attack is a weaker attack that has the following form as an argumentation scheme.

> Person x has argued for thesis A.
>
> But x belongs to or is affiliated with group G.
>
> The views of group G are known to be wrong, or not worth seriously listening to as reasonable arguments.
>
> Therefore, the argument of x for A is wrong, or not worth seriously listening to.

This type of argumentation certainly can be a fallacy as used in some cases. But is it always a fallacy? Or could it be a reasonable kind of argumentation in some cases?

Groarke and Tindale (1986, 310–11) propose that guilt by association can, in some cases, be a reasonable argument, given that practical circumstances may impose limits on how much time one can spend listening to the arguments of this or that group on an issue. They define "good guilt by association reasoning," where someone's arguments are dismissed by virtue of association with a group, under two conditions:

1. he or she really is a member of the group in question, and
2. there is good reason to disregard the views of the group on the subject
   in question, or good reason to denounce the members of the group, and
   no good reason to differentiate the individual in question.

Groarke and Tindale caution, however, that the guilt-by-association argument cannot be used to "definitely" dismiss a person's arguments, because such a definitive dismissal can only be accomplished by a critical discussion of those arguments. Guilt-by-association reasoning, then, is a defeasible argumentation that can be used as a presumptive "shortcut" way of roughly sizing up a person's argument to judge how seriously one should take it or how much discussion time one should take up trying to understand or evaluate it. It is a presumptive reasoning that can go wrong or be used fallaciously in some cases.

A special subtype of the guilt-by-association argument could be called *poisoning the well by alleging group bias*, where a person's argument is attacked by arguing that this person belongs to a group that is biased in favor of its own point of view. In its stronger form, this attack alleges that the group in question has adopted a group quarrel attitude of never really looking at both sides of an issue, of always pushing dogmatically or even fanatically for its own point of view. The allegation, in its strongest form, is that its members are fanatics with whom it is impossible to critically argue or reason. Poisoning the well by alleging group bias, or the PWAGB argument, has the following argumentation scheme.

> Person x has argued for thesis A.
>
> But x belongs to or is affiliated with group G.
>
> It is known that group G is a special-interest partisan group that takes up a biased (dogmatic, prejudiced, fanatical) quarreling attitude in pushing exclusively for its own point of view.
>
> Therefore, one can't engage in open-minded critical discussion of an issue with any members of G, and hence the arguments of x for A are not worth listening to or paying serious attention to in a critical discussion

The PWAGB argument is a strong extension of the guilt-by-association argument, but there are stronger and weaker versions of it, depending on which variant of the third premise is used. In increasing order of strength or severity, it could be argued that the members of G are: (1) biased, (2) dogmatic, (3) prejudiced, or (4) fanatical. The fourth charge is the most severe because it alleges that the group is so hopelessly biased that it is impossible ever to engage in a reasonable and open-minded critical discussion with any of its members. This particular variant is a very strong version of the poisoning-the-well *argumentum ad hominem*. The weaker versions (2) and (3) leave some room for critical discussion. The weakest version is a much milder form of argument that is quite compatible with the critical discussion.

The PWAGB argument can be reasonable in some cases, depending on whether the premises are justified. On the other hand, the PWAGB argument is based on presumptions that may not be justified. If pressed ahead too hard, it can easily be a fallacious instance of the poisoning-the-well *ad hominem* argument. A key problem, noted in the discussion of case 6.10, is that the PWAGB argument, even if it starts out as a weak and justifiable imputation of bias, may eventually polarize the dialogue into two opposed groups, thereby initiating a shift from some more cooperative type of dialogue to a group quarrel.

Threatening Slippery Slopes

The slippery slope argument is a kind of argumentation from negative consequences where a proponent warns a respondent about a dangerous outcome that may occur if the respondent takes a first step in a course of action he or she is contemplating. Every slippery slope argument has three parts. First, there is a first step or initial action that the respondent is considering. Second, there is a series of ensuing steps or consequences that function as additional premises. Third, there is the conclusion, a proposition to the effect that some dangerous outcome or horrible consequence will or might occur through the taking of the first step and the ensuing steps.

Slippery slope arguments can be reasonable arguments in some cases. They are a species of argumentation from consequences used in practical reasoning, especially in advice-giving dialogue. But they can go wrong or become faulty in various ways.

Johnson and Blair (1983, 161) distinguish between a shorter and longer form of the slippery slope argument. In the longer form, a whole series of intervening steps link the initial step and the conclusion. In the shorter form, these intervening steps are omitted, or not explicitly stated as premises, and only the first step and the conclusion are specifically given. Short-form slippery slope arguments are not inherently fallacious, and the missing premises can sometimes be filled in satisfactorily. But they are often based on unexpressed premises, as noted in Jacquette 1989. And because these arguments are so inexplicit, raising a warning about some horrible outcome that might happen, they often take on a powerful force of suggestion, functioning as an indirect threat or scare tactic.

In a case quoted by Johnson and Blair (1983, 164), Canadian unions had been complaining that foreign workers were taking jobs in Canada, in violation of immigration rules. Responding to this worry, the government proposed issuing work permits. Dennis McDermott, who was then the head of the Canadian United Auto Workers union, protested, using the following slippery slope argument.

Case 7.4

They would run counter to our traditional freedoms and would be *the first step* toward a police state.

As Johnson and Blair state, the problem with this short-form argument is that it leaves out all the intervening steps in the supposed slippery slope sequence. Having the power to issue work permits would make it easier for the government to keep track of who works where. And possibly this information could lead to further government restrictions. But is there any real basis for thinking that this would lead to a police state in Canada?

It's hard to answer this question without examining the presupposed sequence of intervening premises in the slippery slope put forward, but one reason why case 7.4 is such a frightening argument is that, leaving so much unstated, it works by innuendo. It is a menacing argument that appeals to fear.

Thus in some cases there is a close connection between the slippery slope argument and the *ad baculum* argument. The element of fear is compounded by the suggestion that once that first step is taken, one is on a slippery slope and has lost all control of the consequences, speeding toward some horrible outcome. Like many *ad baculum* arguments the tactic here uses an indirect speech act. The slippery slope is put forward explicitly as a warning to look out for some negative consequences that might occur if a particular course of action is taken. As such, it can be a perfectly legitimate argument. But the element of menace or scaremongering comes in when the argument is presented in a sketchy but ominous way that suggests some frightening possibility as a conclusive bar to the proposed course of action. Note that in case 7.4 McDermott's argument would have appeared much less menacing if he had used the verb "might" instead of "would."

All slippery slope arguments of the short form, however, are not instances of the *ad baculum* fallacy. Often there is a threatening *ad baculum* element in the use of the slippery slope argument, but this does not mean that the argument is necessarily fallacious. It means that one needs to take a careful look and ask whether those intervening steps can be filled in and justified as premises of the argument.

Reacting to an article in *Newsweek* that was very positive about the trend to manufacture Japanese cars in the United States, Owen Bieber, President of the International Union of United Auto Workers, replied in a letter to the editor ("The Mail," *Newsweek*, 14 May 1990, 8) that the article gave "undue credit" to the Japanese auto makers. According to Bieber, a deeper account of this "transplant phenomenon" should have taken other factors into account, including the fact that "approximately 30 percent of the Japanese nameplate cars sold in the United States are assembled here." Bieber went on to criticize the Japanese practices, citing further statistics.

Case 7.5

> To make matters worse, the Japanese auto firms collectively put less than 40 percent of American parts in their transplant cars, compared to 90 percent for the Big Three. They have shunned traditional U.S. parts makers—including those that meet or exceed their quality and cost standards—preferring to import key components from Japan or buy them from an estimated 250 Japan-based suppliers that have migrated here. This "buying within the family" is destroying a major U.S. manufacturing infrastructure at a cost of tens of thousands of jobs. It threatens to turn the United States into a "branch plant" economy in which we supply basic materials and relatively unskilled labor while many of the most important technologies and higher-order skills remain in Japan.

The conclusion of this argument paints a picture that would be threatening and even frightening to many Americans. The outcome is one in which the United States becomes merely a supplier of basic materials and unskilled labor, while the Japanese manage everything using "higher-order" technical skills that remain in Japan. In effect, the outcome pictured is one where the U.S. becomes an industrial colony of Japan.

This is a slippery slope argumentation that begins by establishing the premise that cars assembled in the United States are made from key Japanese components. The argument goes on to link this first step with the destruction of the manufacturing infrastructure of the United States. Finally, the conclusion is reached that the United States may turn into a "branch plant" economy directed by Japanese skills.

This argument combines the slippery slope with the *ad baculum* through its evocation of a conclusion that would be threatening to Americans. It is not an explicit threat, but indirectly, by evoking a situation that appeals to the emotion of fear, or fearful anticipation, the argument has a key *ad baculum* component in its tactical structure.

Although this case, like the previous one, is an instance of the shorter form of the slippery slope argument, it does back up its premises by appeal to factual considerations, and it does fill in enough of the intervening premises in the slope that it partially fulfills its burden of proof. It would be inaccurate to say that it functions as a simple scare tactic by omitting all the intervening steps from the proposed slippery slope sequence, like case 7.4. And it would not be correct to dismiss case 7.5 as a slippery slope fallacy on the grounds that it claims the threatening conclusion will or must happen, or on the grounds that it gives no evidence to support the missing premises of the intervening series of steps.

But even so, case 7.5 reveals an interesting connection between the slippery slope argument and the type of *ad baculum* argument that appeals to fear. No threat is made by Bieber in case 7.5, but even so, the argument is so powerful precisely because it appeals to a vague fear or unease that is a powerful emotion for the American audience to whom the argument is directed. Possible future consequences are always vague and uncertain, but for this reason arguments that appeal to potentially menacing or disastrous consequences in the future may have a powerful impact of arousing emotions of fear in an audience that may be affected by these consequences.

This case underlines what has been brought out by many other cases, namely, that appeals to emotion are often practical (prudential) arguments directed toward influencing a respondent's actions in the future. As such, they are often based on argumentation from consequences. Also, although they may basically be arguments from consequences, literally interpreted, they may also make indirect appeals to emotions like fear. Such cases require a pragmatic analysis that examines how the argument is being used in a context of dialogue.

Hombac Argumentation

It is possible to have arguments that combine the use of the *ad hominem* and *ad baculum* tactics. A case in point would be an argument where a proponent attacks the character of a respondent on the grounds that the respondent has allegedly been responsible for some reprehensible action and then follows up the argument by threatening to bring this charge to wider public attention (with bad consequences for the respondent) if the respondent fails to carry out some other action. This combined type of argumentation could be called *hombac* argumentation.

Hombac argumentation is not necessarily fallacious. True, hombac argumentation looks pretty reprehensible, perhaps similar to blackmail. Although hombac argumentation has been classified as fallacious by logic textbooks, it is not so simple or straightforward as that. There are instances in which hombac argumentation is used nonfallaciously.

The problem here, as elsewhere, is that the context of dialogue is overlooked. Appeals to threats or fear may not be very nice, and certainly in a critical discussion, threats have no place and are no substitute for reasoned arguments that present evidence or otherwise contribute to the discussion. However, the context of dialogue might be negotiation rather than critical discussion.

Engel (1976) presumes generally in his account of the fallacy of appeal to fear that threats of sanction or embargo by publicizing lists of groups or individuals that support a particular policy or action are fallacious because "the object of the argument is an appeal to emotion rather than reason." He cites the case of the oil-producing countries in the Middle East in 1974 compiling a list of U.S. companies judged to favor Israel (133).

Case 7.6

Appeals to fear tend to multiply during periods of stress or conflict, both among nations and among individuals. When the oil-producing countries of the Middle East assumed a position of threat to the oil supplies of the West in 1974, for example, it became known that they had compiled a list of U.S. companies that they judged to favor Israel. This was interpreted as an appeal to fear directed at American firms that might favor Israel in the Arab-Israeli dispute.

Engel's case makes the good point that during group conflicts and quarrels, emotions do have a way of fueling hostilities and reinforcing the hardening of positions on both sides. But the classification of this case as a fallacious *ad baculum* is based on the erroneous presumption that a listing of companies or individuals judged to engage in or support particular practices is an inherently wrong or fallacious way of supporting an argument.

Consider a parallel case.

Case 7.7

Dolphins tend to swim over large schools of yellow-fin tuna, and when fishermen catch huge quantities of tuna in enormous nets, they also kill many dolphins. The dolphins, as mammals, gasp for air in the snare of the nets, drown while being dragged down, and are even torn apart by the ships' winches while still conscious. When a film showing the dolphins' plight was made public, sentiment against these horrifying practices began to grow. The environmentalist advocates who condemned such practices began listing companies who did, or did not, use "dolphin-friendly" methods of fishing for yellow-fin tuna. They advocated not buying the tuna of companies listed as "non-dolphin-friendly," and due to their campaign, sales of these brands of tuna radically declined in the United States.[1]

1. This case is not a quotation but only a reconstruction from memory of some discourse in a television program.

In this case listing the companies who were engaged in the practice the environmentalists wished to condemn proved very effective. Their campaign depended not only on the threat or "appeal to fear" directed against American firms, but also on an appeal to pity for the plight of the dolphins. There was nothing fallacious about it.

Both these cases also involved the use of the *argumentum ad hominem* because the tactic of attack was—in listing the countries or companies—to identify them with presumably bad actions or positions.

A quite different type of hombac argumentation can be found in case 5.13a, where the president of Dartmouth College charged that the editors of the *Review* had intimidated discourse "by the fear of the kind of personal characterizations that appear in the *Review*." As noted in case 5.13, the *Review* had used *ad hominem* argumentation against selected faculty, calling one a "patron saint" of "little Latin commies" and saying of another, "His courses are filled with fruits, butches and assorted scum of the radical left." The president's allegation was that the editors of the *Review* had used these *ad hominem* arguments as tactics of intimidation, or *ad baculum* arguments used to shut down "uninhibited discourse."

On the other side, the professor who defended the *Review* editors called the charge of racism leveled against them "a slimeball attempt at intimidation." Here too, charges of intimidation were connected with an *ad hominem* attack. Each side accused the other of intimidating the flow of free dialogue ("poisoning the atmosphere") by using "personal characterizations."

In a case like this it is clear that the *ad hominem* arguments are being used in such an excessive and vicious way that they are threatening. But the countercharges of intimidation are difficult to judge fairly, because both sides claim the other side is using unfair tactics of intimidation. At any rate, the case illustrates a species of hombac argumentation, with charges of *ad baculum* and *ad hominem* fallacies being made by both sides to the dispute.

Another aspect of this case is that whatever critical discussion there was supposed to be has evidently degenerated into a group quarrel. Both sides have been so carried away with personal and partisan attacks that the resulting bias and prejudice has left no room at all for real critical discussion. It seems here like bias has gotten the upper hand on both sides and that the argumentation is purely eristic.

Another aspect of case 7.7 is that it not only combines the *ad baculum* and *ad hominem* but also includes the *ad misericordiam*. The film used by the environmentalists, showing dying dolphins gasping for air, is quite clearly an appeal to pity. This case could be called, quite properly then, an instance of *miserhombac argumentation*, which combines three types of emotional appeals.

What must be remembered in evaluating cases like 7.6, however, is that just because an appeal to fear or some other emotion is being used to manipulate someone in argument, it does not follow that the appeal must be an emotional fallacy. It is quite true that "appeals to fear tend to multiply during periods of stress or conflict," as in a group quarrel for example, but that in itself does not necessarily make such an appeal a fallacious argument. In a group quarrel, the *ad baculum*, *ad populum*, *ad hominem*, and *ad misericordiam* are commonly the normal arguments used to carry out the goals of this dialogue. But in such a context there may be no good reason to classify these appeals to emotion as fallacious.

Hompop and Miserhom Argumentation

In Chapter 3 it was shown that political discourse often involves argumentation in a context of several types of dialogue mixed together (see "Political Discourse"). And in Chapter 6 it was shown that a speaker's character is an important factor in many political speeches (see "*Ethos*: The Pedestal Effect"). According to the Aristotelian theory, a political speaker's argument is credible, at least partly because her audience takes her to be a person of good character.[2] On the other hand, in a democracy, a speaker will be perceived as having a good character if she is perceived as a "woman of the people"—a person who has values that many people can identify with.

Putting all this together, a link between the *ad populum* and *ad hominem* is revealed. On Aristotle's theory, character can function as a positive factor that can lead an audience to find an argument more persuasive, convincing, or credible than it would otherwise be. Within an Aristotelian context of deliberative political discourse, this positive evaluation of an argument can be quite correct. In short, the *ad hominem* has a positive side. Though normally viewed as a negative kind of argumentation used to attack and refute someone's argument, the *ad hominem* can be seen as based

2. On Aristotle's theory of political deliberation see Brinton 1986.

on a presumption that a speaker's character bestows a positive credibility on her argument—see Brinton 1985, 1986, 1987.

If the *argumentum ad hominem* is viewed in this positive way, however, the political speaker's character or *ethos* could be the basis of her popular appeal, suggesting a connection with the *argumentum ad populum*.

Ethos is a kind of personal credibility that, according to the Aristotelian theory, is based on a political speaker's experience, practical wisdom, and skills of judgment in deliberation.[3] But there are good questions whether this theory is applicable in modern democratic political discourse. Do people pay attention to a political speaker because of her presumed practical wisdom, skill, or experience? Or is credibility in political discourse more of an *ad populum* appeal to a bond of group solidarity between a charismatic leader and her following? There is a relationship in this area between the *ad hominem* and *ad populum* arguments.

When Lech Walesa, the acclaimed leader of Solidarity, announced his intention to run for president of Poland, it seemed that his popularity as a charismatic leader had started to fade. Opinion polls showed that his "approval rating" slipped. Critics complained that Walesa did not know about the tasks of governing and tended to be unpredictable, one day saying that reforms were too radical and the next day saying that the government was not going fast enough. At a Solidarity union meeting, critics made complaints that suggested Walesa was "losing ground" with the union members.

Case 7.8

> Marian Jurczyk, a breakaway Solidarity leader, accuses him of having "betrayed members' interests in order to pursue personal political ambitions." Some members think success has gone to Walesa's head—and to his expanding waistline. In Gdansk, the cradle of his power, disgruntled workers recently showered him with catcalls like "Fatso" and "Pontoon." They have poked fun at him for riding in a government limousine and wearing a signet ring. ("Doubts about Walesa," *Newsweek*, 23 April 1990, 14)

In this case, Walesa's detractors were using an *ad hominem* attack based on an *ad populum* premise. Not only were they attacking Walesa by arguing

---

3. See "*Ethos*: The Pedestal Effect," Chapter 6, on ethotic argumentation.

that his character, skills of judgment, and personal integrity were questionable and unsuitable for a president, but also, the basis of the attack was the allegation that Walesa had lost touch with the membership, that his popular appeal with the membership had dwindled. It was suggested that his life-style had changed and that he was no longer a "man of the people." One newspaper complained that he was "autocratic" and was "losing popularity" (ibid.). This argument was a kind of reverse *ad populum* attack on Walesa's character and popularity as a union spokesman.

On the other hand, other criticisms of Walesa's character seemed to take the opposite direction, saying he lacked the "personal qualities needed in a president, including culture, knowledge and immaculate manners" (ibid.). Here, Walesa seems to lose either way. If he has nice manners, a signet ring, and other attributes of sophistication, he loses ground with the union. But if he lacks these things, he does not have the qualities needed for a president.

The fact that hompop argumentation is so subtle, complex, and changeable is due to the inherently complex nature of political argumentation as a context of dialogue in a democratic system. The basis of *ethos* is a solidarity of background, character, and point of view between the popular political leader and his electorate. But, on the other hand, the electorate also expects certain skills of good judgment as qualifications for a leader. Any perceived lapse from these popular qualities or skills will immediately be subject to effective *ad hominem* attacks. But specific limits on which aspects of character are judged relevant will vary according to the expectations of the times. The boundaries of political versus purely private matters in political argumentation are subject to changing customs and manners.

A curious example of miserhom argumentation is case 4.11, which combined the *ad misericordiam* and *ad hominem* arguments. Here the American Association for Retired Persons used an appeal to pity, showing an elderly man incapacitated by a stroke, being cared for by his wife. But by also portraying elderly people as active and adventurous in other advertisements, the AARP opened itself to a circumstantial *ad hominem* attack. Thus, the *ad hominem* criticism undercut the appeal to pity by suggesting that the AARP was (illogically) trying to have it both ways. The *ad misericordiam* argument could have been a reasonable appeal, taken by itself. But it was undercut, or brought into question, for *ad hominem* reasons, when placed beside other appeals of the AARP.

Bacpop and Miserpop Argumentation

It is possible for the *ad baculum* and *ad populum* to be combined in a single argument. This combination of argumentation arises commonly in the group quarrel or when there is a shift to a group quarrel. The argument exploits a respondent's fear of exclusion from a popular group. It can take the form of a threat not to include the respondent in the group or its activities, but more often it is an indirect *ad baculum* that appeals in a subtler way to the respondent's fear of social disapproval or of being an outcast.

This combination of the *ad baculum* and *ad populum* arguments, which could be called *bacpop argumentation*, is often a kind of practical reasoning whose aim is to persuade a respondent to undertake a particular course of action. As such, it is a species of negative argumentation from consequences (see Chapter 5, "Argument from Consequences"). The argumentation, on the surface, takes the form of a warning: if you don't do such and such, then negative consequences for you will follow. But beneath the surface, a threat is implied, to the effect that you will be excluded from the (popular) group if you don't perform the action in question.

However some instances of bacpop argumentation are not appeals to a threat. They are, in the typology of figure 5.1, appeals to fear of the nonthreat type. Case 7.9 below is a threat type of bacpop argument, whereas case 7.10 is a nonthreat, or appeal to fear, type of bacpop argument.

Engel cites case 7.9 under the category of "the fallacy of appeal to fear" (1976, 130), calling it a "crude form" of this fallacy because the arguer is explicit about the threat being issued (132). But clearly the case also has elements of the *ad populum* appeal to the feeling of belonging to or identification with the group.

Case 7.9

> You don't want to be a social outcast, do you? Then you'd better join us tomorrow.

This case is similar in certain important respects to case 3.9, the speech by Walter Reuther where the *ad populum* appeal took the form of a "rhetoric of belonging." Reuther's *ad populum* argument placed the value of the group above question by excluding those who wanted to use the movement only "to make a fast buck." In the case 7.9 there is also a kind of reversed use of the *ad populum* that labels anyone not in the group a "social outcast."

Case 7.9, then, is a combination of the *ad populum* and *ad baculum* that appeals to a feeling of group solidarity or exclusiveness while at the same time employing a threat of possible exclusion from this valued group. It is a carrot-and-stick combination of positive and negative appeals. The pain of exclusion depends on and exploits the value of belonging.

Commercial advertisements routinely combine *ad baculum* and *ad populum* tactics of persuasion by appealing to the fear of social disapproval. This appeal is to a fear of not belonging to a popular group, rather than an appeal to threat of physical harm. For example, it may be suggested that if you don't wear a certain brand of jeans, you will be excluded from the fashionable in-group. Or even worse, if you don't use a particular brand of deodorant, you may be smelly and offensive, with the result that your business associates will find you repulsive, and you will find yourself rejected socially, perhaps even failing financially.

Clark (1988) cited the use of fear of social disapproval as a basis for recruitment advertising in World War I. Of the appeals used, one which was alleged to have produced the best results was the following.

Case 7.10

Five questions to men who have *not* enlisted.

1. If you are physically fit and between 19 and 38 years of age, are you really satisfied with what you are doing today?
2. Do you feel happy as you walk along the streets and see *other* men wearing the King's uniform?
3. What will you say in years to come when people ask you— "Where did *you* serve in the Great War?"
4. What will you answer when your children grow up and say, "Father, why weren't you a soldier too?"
5. What would happen to the Empire if every man stayed at home like *you*? (Clark 1988, 112)

Question 2 appeals to envy and/or guilt. Questions 3 through 5 are arguments from negative consequences. But taken together, the questions have presuppositions that target a fear of social disapproval and exclusion in the respondent who has not enlisted.

Case 4.8 was a case of the *argumentum ad misericordiam* where a speaker in a parliamentary debate appealed to pity on behalf of the tobacco farmers, a group who had suffered economic losses resulting in "bankruptcies, family breakdowns and suicides." But in this same speech it is possible to see a heavy strain of *ad populum* appeal mixed in with the *ad misericordiam* argumentation. Mr. Bradley, in his speech, described the tobacco farmers as "people who over generations have developed blow sand into soil and developed a healthy plant from a weed," adding that "[i]n most cases they were immigrants who became the proudest of Canadians." Now, according to Mr. Bradley, society is "shunning" these tobacco farmers, and they have become "outcasts." In conclusion, Mr. Bradley urged his audience: "do not turn your backs on this group of proud Canadians."

This speech has as its main thread of argumentation a curious blend of the *argumentum ad misericordiam* and the *argumentum ad baculum* and is therefore a case of miserpop argumentation. Mr. Bradley is appealing to his audience's solidarity with the tobacco farmers as fellow Canadians, urging his audience not to turn their backs on this pitiable group. In other cases, the *ad populum* was used as a dividing tactic to exclude those not in the group—see case 3.9, for example. Here it was combined with an appeal to pity as a plea not to exclude some members of a group.

It used to be thought that these cases that combine more than one type of traditional fallacy are highly problematic (see Walton 1987b). If such borderline cases exist, how can one divide one fallacy from the other, or tell the difference between the two?

To be sure, such borderline cases are puzzles. But the foregoing cases have demonstrated that they are not the big problem they appeared to be, and they can often be resolved or explained quite satisfactorily once an analysis of each of the individual fallacies has been given. Previously the problem had been that a clear enough analysis of these fallacies was not available.

According to the foregoing analyses of these four fallacies of emotional appeal, each of them needs to be seen, first and foremost, as an argumentation tactic that can be used, properly or improperly, in a context of dialogue. Given this analysis, it is no longer surprising that the same type of argument can be used fallaciously in one case and quite reasonably, or nonfallaciously, in another case. It should also no longer be surprising that two of these tactics of argumentation can be combined in the same case, one reinforcing the other. Even three can be combined or woven together in the argumentation in the same case.

8

RIGHT AND WRONG USE OF
EMOTIONAL APPEALS

In all of the cases examined there is a given climate of opinion, a *prima facie* argument on one side of an issue where there are open questions or practical but nonconclusive reasons why a weight of presumption lies on one side. And in each case there exists doubt, so that the argument hangs in the balance, even if the weight of presumption on one side is heavier than that on the other. Existing knowledge is insufficient to tilt the balance decisively to one side, leaving no room for argument at all on the other side.

In this situation, the four appeals to emotion are used to take advantage of the equilibrium by introducing just enough of a factor to shift the balance toward one side. For example, in case 4.5, the excuse offered might be a weak one, but nevertheless, if it is a borderline case and room for judgment is needed in weighing its seriousness, perhaps just a little push further, supplied by the appeal to pity, might be enough to tilt the scales toward accepting the excuse.

The *ad misericordiam*, the *ad baculum*, and the *ad populum* are all techniques that are aimed toward some bias, emotional predisposition, or susceptibility of the respondent in dialogue that can be appealed to, in order to provide the little shift needed to move the balance. The *ad hominem* is used to convince a third party audience that one's opponent in dialogue is dishonest or insincere in a way that detracts from the weight of presumption accorded to the opponent's argument. The *ad populum* shares this three-person dialectical structure as well, in a more implicit way, by appealing to the accepted climate of opinion of a group, a third party in the framework of dialogue.

In any argument where burden of proof is involved and the case hangs in a balance, there is always an existing bias on the part of the respondent whom the proponent is trying to convince. The respondent has some special interest in the one side of the issue or some emotional pull that inclines her toward this side. This inclination is typically based on practical matters like the respondent's concern for her personal safety, her belonging to some group that she strongly identifies with, or her feeling of sympathy for someone who is in need of help. Appealing to one of these emotional inclinations is therefore always a handy tool that can be used at the right juncture in argumentation with productive effect.

As tools, these arguments are neither right nor wrong, in themselves. But since they can be used wrongly, as well as rightly, in a given case, one needs to be on guard against them. They are powerful tools in some cases, and therefore it is useful to know how they work, both to guard against them and to use them yourself.

Value of Arguments That Appeal to Emotion

Studying the four fallacies of emotional appeal reveals that they are based on kinds of argumentation that are, in some cases, reasonable and nonfallacious. Although there are justifications for the traditional mistrust of emotional appeals in argumentation, those appeals can also be good presumptive guides to action or to a provisional conclusion in opinion-based dialogue. But if these emotional appeals can be reasonable kinds of argumentation in some cases, what is meant here by "reasonable"? What goals of dialogue do they contribute to? What is the function of an appeal to emotion in argumentation? These questions made no sense, viewing good reasoning in the semantic framework of truth-conditions and deductive validity. But in a pragma-dialectical framework, they need to be asked.

What is the positive value of an appeal to emotion in argumentation? Boldly expressed, this question seems strange at first, given the tradition of denigrating appeals to emotion as fallacious. But given that emotional appeals can be reasonable arguments in some cases, the question naturally arises whether, as a class of arguments, they have some value or constructive function in reasoned dialogue.

To answer this question, one must look at different functions of argument in different types of dialogue. Clearly one of the most important types is the critical discussion, but the critical discussion is a type of persuasion dialogue, so the latter is a good place to begin.

The basis of all persuasive argumentation lies in the choice of suitable initial premises for convincing your respondent through your ability to have empathy with your respondent. Empathy is the ability to put yourself inside your opponent's position in an argument, metaphorically speaking—it is the ability to arrive at presumptive conclusions concerning your respondent's commitments in a dialogue. It is based on presumption because, characteristically, you do not have any direct way of knowing what your respondent's commitments really are in an argumentative exchange. You can only partially judge these things by seeing how your respondent reacts to questions and suggestions and by judging what he seems to place most importance on.

Of course, you can determine your respondent's explicit commitments by asking him questions or by paying attention to his explicit assertions. But to construct a truly persuasive argument you must go beyond this stage and form conjectures concerning the deeper but unarticulated commitments on which these explicit commitments are based. These implicit commitments are called dark-side commitments because they are not explicitly known or stated in a dialogue, but nevertheless they exist.[1] But they can only be identified by a process of inference and presumption, using empathy.

How can you tell what another person is feeling? You can't feel what they are feeling. It's something you have to guess at. You have feelings, and you presume that they have the same feelings as you.

1. On dark-side commitments see Walton 1987b (125).

Pity is connected to solidarity with popular opinion because both of these emotions are based on empathic feelings of identification with the position of another person or persons. Pity is also connected to fear; you pity another person in an unfortunate situation because you have a fearful response to the empathic possibility that you yourself could be in that situation, suffering the same consequences. Thus the *ad baculum, ad misericordiam,* and *ad populum* are all appeals that involve an empathy of one person toward another person or persons. There is a kind of solidarity or sympathetic emotional response that transfers from one person to the other. Because you are sympathetic with another person's presumed emotions, there is a bonding, and you are affected by similar emotions that produce a response like pity, fear, or identification with a common position.

The *ad hominem* appeal also involves empathy, because it is a personalization of an argument where one arguer has to try to reconstruct the personal commitments of the other. The *ad hominem* is a negative argumentation used to attack and refute a person's argument, whereas *ad baculum, ad misericordiam,* and *ad populum* are all positive arguments used to get someone to accept a conclusion. But like these other three types of argument, the *ad hominem* is based on an empathic emotional response, because the focus is on the character and presumed deeper personal commitments of the arguer.

Emotions have a steering function as presumptive bases for reasoning. In the inquiry, a knowledge-based type of dialogue, appeals to emotion have no place. But practical reasoning is guided by intelligence based on variable circumstances. And the critical discussion is an opinion-based kind of argumentation. Hence, presumptive reasoning is characteristic of prudential argumentation leading to conclusions on how to act, and of critical discussions to resolve a conflict of opinions on controversial issues. In such contexts of dialogue, appeals to emotion have a presumptive status as provisional bases for steering a prudential course of action or for coming to a critically argued conclusion that reflects a reasoned opinion.

The four emotional appeals considered here guide actions on the following presumptive bases or maxims.

If it is fearful, steer away from it.

If it accords with popular opinion, steer along with it.

If you feel sympathy, steer towards supportive action.

If your informant appears to have bad character (particularly for veracity), steer away from accepting what he says as true or taking what he says at face value.

There is nothing wrong or fallacious *per se* with appeals to emotion in argumentation. Emotion should not be (categorically) opposed to reason, even though appeals to emotion can go wrong or be exploited in some cases.

Heeding your emotions in argumentation can, in general, be a good guide to keeping in harmony with your deepest, fundamental commitments, which define your personal stance or considered judgment as an individual.

Each one of us has personal biases. *Bias* is an initial proclivity to accept some presumptions and reject others on the basis of one's personal interests, circumstances, commitments, and point of view (see "Bias *Ad Hominem*," Chapter 6). Bias steers you one way or the other in a conflict of opinions in practical reasoning or opinion-based argumentation. Where knowledge is insufficient to conclusively resolve a conflict of opinions, bias tilts a burden of proof one way or the other.

Emotional appeals work by appealing to a respondent's strongest personal bias. But not all bias is bad bias. In a critical discussion, one's personal biases can be identified, tested, and evaluated. One can "learn," or advance toward knowledge, through becoming aware of a personal bias. Through a successful critical discussion, one can become less dogmatic and prejudiced by raising appropriate critical questions. The critical function of argumentation can therefore act as a corrective or balance to the steering function of emotions.

The most general maxim is provisionally to follow your instincts, but subject them to critical questioning when time and circumstances for reflection and discussion permit second thoughts. One's considered judgment on a matter of opinion, therefore, should be based on emotion but balanced by criticism.

One's dark-side commitments are based on deep feelings and emotions. A stimulation of the passions can lead to an emotional crisis, and then these emotions can burst out in a quarrel. If the quarrel has a good cathartic effect, a feeling of relief follows the airing of grievances. The emotions are released, and to some extent one's deeper feelings of what is significant may be expressed. But expression of feelings in the quarrel is always a temporary satisfaction.

In a good critical discussion it is possible to get a more lasting feeling of release by actually expressing these feelings in a cognitive form. The dark-side commitments are expressed as explicit commitments and defended and sharpened in the fires of argument. This is not a cathartic effect of release so much as it is an effect of satisfaction at stating and defending one's deeply held commitments, the values that one feels are central to oneself as a person, a distinctive individual with an upbringing and cultural background that has shaped a point of view.

The maieutic function is not simply a release of emotions but a verbal expression of one's deeply held feelings in a form shaped by the rules and procedures of a critical discussion. This verbal articulation in argument of one's dark-side commitments enables them to be partly revealed to oneself and at the same time to others. It represents a critically structured expression of personal insight.

Four Kinds of Argument Defined

In defining each of the four arguments *ad*, classified as emotional in nature, it is structurally significant to note a key difference between the *ad hominem* and the other three. In the cases of the *ad baculum, ad misericordiam,* and *ad populum,* the argument is directed by one participant in a dialogue towards the emotional susceptibility of the other participant—it is an appeal *to* the other participant, designed to appeal to some commitment or tendency of that participant, to get him or her to accept a conclusion by using the appeal in a positive way. In the case of the *ad hominem,* the appeal is directed by the first participant towards a third-party audience (judge, respondent, etc.) in order to get that audience to discount, or not to accept, the argument of a second party. The reason given is that this second party is an unreliable source, because of his or her bad character for veracity or reliability as a participant in collaborative dialogue. This is a different dialectical structure from that of the other three arguments *ad.*

This structural difference comes out when the definitions of all four types of arguments *ad* are given. Below are the definitions that have emerged from the detailed analyses of the case studies.

The *argumentum ad baculum* is a kind of argument used in dialogue to shift a weight of presumption toward one's conclusion by appealing to one's respondent's timidity or fear, especially by conveying (overtly or covertly) a threat to that respondent's interests or safety. ·

The *argumentum ad populum* is a kind of argument used in dialogue to shift a weight of presumption toward one's conclusion by appealing to the respondent's susceptibility to go along with the commitments of some group he or she belongs to and in particular to accept the received popular opinion on an issue. The emotion of belonging is here appealed to by exploiting the respondent's longing to conform to popularly accepted and approved practices, especially if such a social alignment to the popular group values or practices has an existing presumption in favor of it.

The *argumentum ad misericordiam* is a kind of argument used in dialogue to shift a weight of presumption toward one's conclusion by appealing to a respondent's feeling of pity or sympathy for some person who is thought to be in need of help and who could presumably be given that help by the respondent's assent to the conclusion at issue. The *ad misericordiam* is an appeal to the fundamental empathic responses, the human need to help another person who is in a difficult or painful situation. Like the other two arguments *ad*, it is clearly an appeal to emotion that is the mainspring of the argument.

With the *argumentum ad hominem*, the situation is quite different. Here, one arguer is attacking another arguer negatively, trying to run down or refute her argument by "badmouthing" her. This is not an appeal to some emotion that will positively support the argument for the first arguer's case. Instead, the first arguer is trying to run down or rebut the argument of the second arguer (in the eyes of a third party) by denigrating the character of the second arguer. Accordingly, the *argumentum ad hominem* is best defined as a kind of argument used to shift a weight of presumption toward one's own side of a disputed question by attacking the character of the participant in dialogue who has supported the other side. Such an attack can be carried out in various ways—by attacking the person's character directly, especially her character for veracity, by attacking her personal consistency of commitments on the issue, or by alleging that she has a personal bias that throws her credibility into question.

Given this essential difference between the *ad hominem* and the other three types of appeal to emotion, a critic might ask why the *ad hominem* is classified under the category of emotional fallacies at all. Isn't it too different from the other three in how it works?

This objection has a good point at its basis. As an argument, the *ad hominem* perhaps does not belong in the same category of argument based on appeal to emotion as the first three types of argument. However, as a fallacy, the *argumentum ad hominem* definitely belongs to the category of emotional fallacies, as the cases in Chapter 6 amply demonstrate. The key thing about the *ad hominem*, as a fallacy, as noted in here, is that it functions so effectively as a sophistical tactic precisely because it brings the argument to a personal level, along with all heightened emotions that are brought along with this focus. Most notably, the *ad hominem* is such a notorious and powerful fallacy because it tends to prompt, and is associated with, the dialectical shift from some other type of dialogue to a personal or group quarrel. For this reason, the *ad hominem* definitely belongs with the other three under the heading of emotional fallacies.

Emotional Fallacies

In many of the cases of arguments based on the four kinds of emotional appeals examined in the foregoing chapters, the powerful appeal to an emotion was used to get an audience to accept a prejudged, one-sided, or biased point of view without looking too carefully at relevant evidence in a more balanced way. These were weak arguments, however, not fallacious arguments. The fallacious cases were characterized by a hardening of bias into a dogmatic, prejudiced, or even overzealous or fanatical attitude of pushing the emotional appeal aggressively ahead, trying to make it bear much more weight than it ever could in a reasoned dialogue. In these fallacious cases, the emotional appeal was used to suppress all possible critical questioning by a respondent, in effect closing down the sequence of legitimate dialogue that was supposed to take place.

In the fallacious cases, the tactics used expressed a closed attitude on the part of the participant in dialogue who committed the fallacy, at least insofar as such attitudes can be judged by the argumentation that is their expression, according to the given evidence of the text and context of dialogue in a particular case.[2]

A fallacy is a violation of a rule of reasonable dialogue. But it is more than that as well. It is the use of a tactic to block or subvert the legitimate goals of a dialogue the participants are supposed to be engaged in. With the

2. On attitudes appropriate for collaborative speech acts in argumentation in a critical discussion, see van Eemeren and Grootendorst (1984).

emotional fallacies, uses of such tactics are associated with a hardened bias that leaves no room for free and proper dialogue.

It was concluded in the last section of Chapter 3 that the *argumentum ad populum* becomes a fallacy when the presumption of popular opinion as a guide to acceptance of a proposition is advanced by a proponent to preempt or block a respondent's critical questioning, thus shutting down further avenues of discussion. This tactic of hindering the legitimate goals of a dialogue is in fact generally characteristic of a dogmatic, prejudiced, or even fanatical attitude. It is one thing to commit the error or blunder of putting a popular presumption forward without responding to the need or request to back it up with evidence or to prove it. It is another thing to commit an *ad populum* fallacy by expressing an attitude that shows strong dialectical evidence of a refusal to allow such critical questions to be asked or that systematically blocks or discounts all attempts to ask such questions. Such a hardened, closed attitude, characteristic of the dogmatic or prejudiced arguer, is rightly to be equated with the committing of the *ad populum* fallacy.

Cases 3.6 through 3.9 are therefore rightly classified as instances of the *ad populum* fallacy. In all of these cases there is clear textual evidence of the proponent's preemptive tactic of setting the conclusion so firmly in place that any further critical questioning of it is overridden or closed off.

The problem with many *ad misericordiam* arguments is that the appeal to pity has a tremendously moving impact on the audience, which may lead them to overlook that not enough relevant facts have been given to judge the case fairly. For example, in case 4.12 the emotional response aroused by the graphic description of a grief-stricken mother clutching her two-year-old son as they bid a tearful farewell to his father is quite likely to leave the casual reader with a one-sided or prejudiced viewpoint on the case. Critical questions concerning the specific reasons behind the deportation of the father tend to get left behind or overlooked. The problem with such persuasive *ad misericordiam* appeals is that by appealing to the emotions of the audience, they may get by with presenting a biased account that goes unquestioned.

However, as noted in Chapter 4, many cases of *ad misericordiam* argumentation are better classified as weak arguments rather than fallacies, for a biased argument is not necessarily a fallacious argument.

Where the *ad misericordiam* argument is fallacious, as noted in the last section of Chapter 4, it has been used to prevent critical questioning by pressing ahead too aggressively in a dialogue, trading on the decisive impact of a powerful appeal to pity. It is exactly in this type of case that the proponent tends to be dogmatic or prejudiced, rather than simply biased, as shown by the use of tactics to block off the possibility of legitimate critical questioning by a respondent audience.

The *ad misericordiam* can also be designed to force a dialectical shift.

Case 8.1

> A student approaches a professor to ask that she should be given a higher grade on her test. She cannot point to any specific test items when the professor asks her why she should be given a higher grade, and instead replies: "I need an "A" average to get into law school, and if I don't get an "A" on this test, I will not get an "A" average."[3]

The *ad misericordiam* appeal is used to shift the dialogue from a discussion of whether any of the test answers merit a higher grade to argumentation from consequences concerning failure to gain entry to law school. The *ad misericordiam* appeal is fallacious in this case because it is used to put the professor in a position where he is to make the decision in a biased way.

In this case, the professor is supposed to judge the test according to the academic and professional standards of his field of expertise. But the student is proposing that the judgment be made on an *ad misericordiam* basis of the consequence of her not getting into law school. This tactic is an attempt to limit the proper freedom of the professor to express his academic and professional standpoint by judging a grade by the standards he should use in this type of case.

In this case the *ad misericordiam* argument is fallacious because there is a dialectical shift from a given type of dialogue the participants are supposed to be engaged in. The student and the professor are supposed to be engaged in a pedagogical dialogue where the professor is testing the student by asking examination questions. The student has the obligation of answering these questions in a way that is supposed to show the professor she has learned something about the subject. When the professor grades the answers, he is supposed to do so in a fair and professional way, so that the

---

3. This example is due to Anthony Blair, who brought it forward to illustrate a point following a panel discussion on the fallacies on June 29, 1990, during the Summer Institute on Argumentation at the University of Amsterdam. In evaluating case 8.1, the reader should turn back to other cases in the "Excuses" section of chapter 4 that occurred in a similar context.

grades given reflect the real merit of the answers in the field of expertise. He is not supposed to give grades in a biased manner, or for reasons other than the comparative academic merits of the answers. This is the context of dialogue in which arguments are evaluated.

When the student replies, "I need an 'A' to get into law school," this remark could be quite true. But as an argument or part of an argument from consequences that is an appeal to pity, it needs to be evaluated according to the rules and requirements of the original context of dialogue that the student and the professor are supposed to be engaged in. From this standard, it falls short. More than that, from this standard, it is an irrelevant argument. And even further, it is a kind of argumentation that shifts the context of dialogue away from the original type of dialogue the participants are supposed to be engaged in.

The *argumentum ad baculum* is a fallacy when a proponent in a critical discussion puts pressure on a respondent by making a threat (or fearful appeal) in order to prevent that respondent from questioning the proponent's standpoint. Hence this fallacy, too, expresses a prejudicial or dogmatic attitude on the part of the arguer who commits it by inhibiting critical questioning.

In case 5.11, for example, the speaker puts forward a proposal at a business meeting but then threatens to "curtail the operations" of anyone who might raise critical questions about the proposal. Anyone who opposes the proposal will invite retribution, and so the upshot of the threat is to close off any critical discussion of the proposal by the audience. The speaker's attitude is dogmatic or prejudicial; he steamrollers ahead with the proposal, making it plain that he is not prepared to listen to any critical questioning or open discussion of the merits of his proposal.

In case 5.14 the indirect threat posed by the Direct Action terrorist, concealed in the question about security measures, had the same effect of closing down the avenues for subsequent dialogue on his case during the trial procedure that was supposed to take place. It would not be improper to say that the *ad baculum* in this case is associated with a fanatical attitude that leaves no room for real critical discussion as a way of resolving a conflict of opinions.

The *ad hominem* is an attack on the reliability of a source of information or argumentation. The circumstantial *ad hominem* alleges that an arguer's personal circumstances are inconsistent with his argumentation, suggesting that he is insincere as a reliable or cooperative participant in reasonable dialogue. The bias type of *ad hominem* alleges that an arguer has a personal bias, suggesting that his impartiality as a source cannot be taken for granted and thereby implying that his argument ought to be discounted. Many bias *ad hominem* arguments are quite reasonable if interpreted as presumptive arguments that raise critical questions about the credibility of a source.

Sometimes the bias and circumstantial *ad hominem* arguments are combined, as in case 6.6, where Juan Ponce Enrile accused Jose Cojuangco, Jr., of being a "private money machine for Marcos cronies." The problem was that Enrile had also had a long career as a Marcos ally, during which he had prospered in various industries. Noting the circumstantial inconsistency, *Newsweek* commented on the "partisan tinge" of Enrile's argument. Here the circumstantial inconsistency combined with a suggestion of bias due to financial interests, resulting in a double-barreled *ad hominem* attack.

The *ad hominem* becomes problematic, however, when the allegation of bias or personal attack is based on or leads to a shift to a quarrel. For example, the *ad hominem* bias allegation in case 7.3 prepared the way for an onset of the group quarrel by trying to divide the dispute into two opposed sides—men and women—alleging that it is "impossible" for men to see the issue from the woman's point of view.

Emotional Appeal and Bias

Appeals to emotion in argumentation become fallacious not simply because they are invalid or erroneous inferences, but because they are used to capitalize on a bias that shifts or twists the context of dialogue. They are used in dialogue to appeal to the existing bias of a respondent in order to steer the respondent away from fulfilling her obligation in the dialogue to argue or ask questions according to the rules of dialogue that the participants are supposed to be engaged in. Instead, the respondent is invited to succumb to the pull of her vested interests or strongest biases.

At the same time, the proponent conveys a biased and antilogical attitude by using a fallacious emotional appeal to limit the freedom of the respondent to raise the right critical questions or to properly express her standpoint. Normatively, there should be a free flow of dialogue, so that a respondent can reply to an argument in whatever way she thinks will best fulfill her obligations or express her views. An emotional fallacy is a snare and deception that puts a respondent in a position where it becomes difficult for her to judge the issue according to the standards she should use, or is supposed to use, in order to properly decide on the issue.

However, a fallacy is not necessarily an *intentional* deception or trick to trap an opponent. But the use of a fallacy in a given case often (and in the case of the four emotional fallacies, characteristically) conveys an attitude on the part of the proponent of not being open or sympathetic to the proper aims and methods of the dialogue he is supposed to be taking part in. There is a failure here to live up to Gricean maxims of honesty and helpfulness in contributing to the dialogue. By trying to appeal to the respondent's bias and by using subversive tactics to go against the goals of a dialogue, the proponent who commits an emotional fallacy exhibits an attitude of hardened bias, pushing too aggressively for a one-sided point of view.

Bias is not all bad bias, but the exploitation of bias through the use of powerful emotional appeals is something to watch out for in argumentation. In the section on "Types of Bias" below, two kinds of bias are identified. One is caused by the use of different functions of argument at cross-purposes in dialogue. The other is caused by a dialectical shift from one type of dialogue to another.

Defining Bias

Defining bias is a tricky business, as case 6.7 has already indicated. In this case, Bob accuses Wilma of bias on the grounds of her financial affiliation with a large U.S. coal company. Whether this accusation of bias is justified or not, as a critical questioning or refutation of Wilma's argument, depends on the context of dialogue. If the dialogue is supposed to be an impartial inquiry, Bob's criticism that Wilma's bias is an *ad hominem* fallacy in her argument could be fully justified. But even if the dialogue is supposed to be a critical discussion, where Wilma supports the side of the industrial companies, her failure to announce her affiliation would be rightly regarded as an important criticism of her argumentation, once Bob announces it.

Judging from this case, it seems that bias is not simply a failure of neutrality, but rather a failure of the kind of balance of neutrality appropriate for the type of dialogue the participants are supposed to be engaged in. Wilma is not supposed to be completely neutral or dispassionate in her critical discussion of the issue of acid rain. But if she has concealed important and relevant financial interests in the outcome, it may be that she is covertly engaged in a negotiation or interest-based type of dialogue.

Two essential characteristics define bias. One is lack of balance or neutrality in argumentation. The other is lack of critical doubt in argumentation. It is easy to confuse these two aspects, especially if critical doubt is defined as a zero or neutral point of view in an argument, that is, neither *pro* nor *contra*. But following van Eemeren and Grootendorst (1984, 81), critical doubt is an attitude toward the attitude (point of view) of another participant engaged in argumentation. So defined, critical doubt is a normative concept, because it entails taking up a proper or appropriate questioning stance against the expressed point of view of the other party engaged in a critical discussion.

Lack of balance or neutrality is not always a problem or failure in argumentation. For example, if the dialogue context is a quarrel, then it will be supposed that one side will passionately support its view and attack the opponent's side, showing lack of neutrality. This may also occur in a critical discussion, and it is not a problem or critical failure.

In other cases, however, lack of balance or neutrality can be a serious failure. Suppose a reporter, who is supposed to give news coverage of a controversial issue, continually supports one side and belittles or runs down the point of view of the other side. Such a display of bias could be a journalistic and critical failure.

Or suppose in a critical discussion you ignore obvious critical weaknesses in your own argument and try to cover them up with passionate emotional appeals that don't respond to the critical questions posed. This kind of display of bias is a critical failure.

Thus bias, when it is a harmful bias or critical failing in argumentation, is a lack of balance or critical doubt appropriate for the context of dialogue an arguer is supposed to be engaged in. This means that the same passionate argument could be biased, in the sense of being a critical failure, in one context of dialogue, yet not biased, in the sense of being a critical failure, in another context of dialogue. Yet it is the same argument (apart from its context). This idea is simply a consequence of defining bias as a pragmatic and dialectical concept. Whether an argument is biased (in a way that should make it subject to criticism) is a function of how that argument is used in a context of dialogue.

To return to the case of Wilma and Bob (case 6.7), Bob's *ad hominem* criticism of Wilma's argumentation has legitimate impact as a criticism not because of any of Wilma's individual arguments, in themselves. Wilma's alleged bias is a critical problem because it appears that she has been dishonest and insincere, thereby failing to meet the Gricean maxims for polite collaboration in the critical discussion on acid rain. The alleged facts brought forward by Bob raise questions about Wilma's sincerity in putting forward the arguments she does. It appears that covertly Wilma may have been engaging in a different kind of dialogue all along.

This does not mean Wilma's arguments are wrong, in themselves, that is, invalid or based on false premises. On the complex and controversial issue of acid rain, truth is hard to come by, and knowledge or hard evidence does not seem to resolve the conflict of opinions. Given the openness of the issue, whom is one to believe? Not Wilma perhaps, if she is really pushing for the interests of the industrialists. The room for doubt in this case swings a burden of skepticism against Wilma's side.

What is most important about bias, then, is that, as a critical failure of argumentation, it is more than just a lack of neutrality. It is a lack of balance or neutrality appropriate for the type of dialogue the arguer is supposed to be engaged in. As such, it is a critical failure because it is a failure of the functioning of critical doubt.

Critical Doubt

In a quarrel, the attitude that will eventually contribute to the resolution of the quarrel is a strong *pro* attitude that opens up one's deepest feelings about why one is in the right. But the critical discussion is a more complicated dialogue that requires different attitudes for different argument situations. First, a critical discussion is adversarial, and it does help the progress of the discussion if the protagonist of a point of view pushes ahead strongly to support that viewpoint positively by marshaling the strongest arguments in favor of it that she can find. But pressing ahead strongly, without respite or reconsideration in every situation, can, in some instances, actually be a factor that blocks or hinders the resolution of the conflict of opinions.

Instead of pressing ahead blindly, in some situations a proponent must sit back, so to speak, and bracket or suspend this forward-moving *pro* attitude temporarily. Taking a "critic's-eye" point of view, an arguer must sometimes try to look at her own argument from her opponent's point of view, asking, "What are the weakest links in my argumentation, from his point of view?"

This anticipatory view of possible objections, however, requires a different attitude from simple *pro* argumentation. It involves a suspension of the *pro* attitude, at least temporarily, and an ability to look at the issue from the other side—empathy.

Another attitude that is important for a critical discussion and that requires a temporary suspension of either *pro* or *contra* attitudes is the attitude of critical doubt. Some might define critical doubt as a neutral or zero attitude, one that is neither *pro* nor *contra*. This way of defining critical doubt sees it as emotional detachment.

It is often said that in order to deal with appeals to emotion in argumentation, one needs to be calm, dispassionate, or detached. According to Thouless (1930, 232), it is difficult to determine right opinions on a disputed subject because of the strength of our "hidden emotional inclinations" on that subject. To contend with such bias, Thouless (232) recommended the cultivation of an "attitude of detachment of mind." But detachment from emotions is not always a good thing in argumentation. In many instances, in order to make a convincing case, it is important to show a passionate conviction. Not every presumption or bias is harmful to critical thinking. The problem is to judge when a passionate inclination has become an interfering bias, a critical failure that is an obstacle to good reasoning.

Instead of seeing critical doubt as an attitude of dispassionate detachment that excludes or even suppresses emotions, it could be seen as a second-order attitude, an attitude toward an attitude in a critical discussion. This is the way van Eemeren and Grootendorst (1984, 81) define critical doubt:

> It is important to realize that the doubt expressed by a language user in a dispute does not bear directly on the expressed opinion but on the *point of view* or *attitude* expressed by another language user *in respect of* the expressed opinion. Perhaps it is also important here to observe once more that expressing doubt, while it may *accompany* the adoption of the opposite attitude, is *not identical* to propounding the opposite point of view.

In a critical discussion, according to van Eemeren and Grootendorst's account, two parties have set out to resolve an externalized conflict of opinions, and each party has a point of view (standpoint). A standpoint has two components: (1) a proposition, representing the thesis (conclusion) a party is arguing for, and (2) an attitude toward this proposition. An attitude can be positive, negative, or neutral (79). Critical doubt, according to the account given in the quotation above, is an attitude of one party in a dispute toward the attitude of the other party.

How does this idea of critical doubt work in practice? Consider the case of a critical discussion between two parties on a controversial issue where the two parties have opposed passionate convictions. Neither party is neutral, and both may be said to be biased in favor of his or her own point of view.

In such a case, critical doubt is not having a neutral point of view. It is the bracketing or suspending of the point of view you already have, in order to express doubts and questions. But such a suspension does not necessarily imply a neutral attitude. Instead, it means that you retain your positive attitude in favor of your own point of view, but in addition, you superimpose a second-order attitude upon that first-order attitude.

The concept of critical doubt, so conceived, is a subtle one. It requires that an arguer must be able to play two distinct roles in a dialogue at the same time. She must push ahead with her strongest arguments in favor of her point of view, yet at other points in the same dialogue she must maintain this positive point of view while adopting the second-order attitude of evaluating that argument to see where appropriate critical questions need to be asked.

To perform this function successfully, a participant in argumentation must resist the natural impulse to press ahead aggressively with the partisan role of arguing forcefully for her own point of view. The dogmatic or inflexible arguer is unable to carry out this function well. Such an arguer tends to see her opponent as being the dogmatic or fanatical one—a person who is so wrapped up in expounding his own point of view that he is unable to rationally evaluate an argument because of his personal bias.

It is in just this kind of case that the critical discussion tends to focus on personal attack on both sides. The problem in such a case is that the critical discussion deteriorates into a quarrelsome dialogue. This is the type of situation where fallacies tend to be committed, precisely because the quarrel leaves no room for the functioning of critical doubt necessary for a successful critical discussion to take place. The one party tends to presume that the other party is in the wrong, showing no respect for the capability of the other party to recognize a good argument. She therefore feels justified in attacking the other party, using *ad hominem* argumentation. She portrays him in her arguments as a person who has no regard for the truth.

Each party then tries to browbeat the other party with aggressive and dogmatic appeals to expert opinion and other tactics. These combative tactics, which might in other cases not be wrong or inappropriate to use in a critical discussion, can nevertheless be pushed forward in such a heavy-handed, one-sided, and aggressive way that they become serious obstacles to the continuation of dialogue. Once both parties give in to participating unrestrainedly in this quarreling exchange, the reasoned discussion of the issue becomes hopelessly blocked.

Types of Bias

Bias is relative to what a dialogue purports to be. If a dialogue purports to be a partisan defense of a particular point of view, an argument occurring in it may not be biased, in that context of dialogue. But the very same argument may be biased were it to occur in the context of what purports to be an impartial discussion of the issue.

Sometimes bias is not a problem—it does not interfere with the goals of a dialogue. Blair (1988) distinguishes between "good bias" and "bad bias." Bad bias may take the form of prejudice, dogmatism, or even fanaticism, in extreme cases. What is meant by "bad" here is functionally bad, in the sense of preventing a dialogue from functioning smoothly or properly.

A judge or mediator of an argument is supposed to be impartial. Therefore, any bias shown will be rightly taken to be bad bias. However, an active participant in the argument will be expected to show bias in actively supporting his own side of the argument. There may be nothing critically wrong with this type of bias, and it could be called "good bias."

In a critical discussion, the dialogue should be a critical test of the point of view under discussion. A certain degree of bias does not prevent this, but bias can reach a point where it interferes. Bias interferes in two types of cases. One is where an arguer becomes so emotionally involved in strongly advocating his own position that there is a loss of the capability to see an argument from any other point of view. This is a loss of critical detachment or empathy within the dialogue. The other kind of case where bias interferes concerns an illicit dialectical shift where an arguer is no longer sincerely taking part in the dialogue he is supposed to be engaging in but is instead taking part in some other type of dialogue, whether covertly or unknowingly. In such a case, it may seem (to one party or the other, or both) that the party is engaging in the dialogue, but really he is not. This disparity between what seems to be the case and what really is the case can be explained as a dialectical shift.

Internal bad bias arises from an improper mixing of three functions or skills of argumentation in dialogue. The *supportive function* is that of defending your own position from your own point of view and expressing this point of view in argumentation. The *empathetic function* is putting yourself "inside" your opponent's point of view, to see how to argue with him convincingly and effectively. This means persuading your opponent by arguing from what you presume are his commitments. This skill requires appreciating your opponent's point of view and involves the ability to "step outside" of your own point of view. The *critical function* is taking a stance of critical doubt by assessing the strengths and weaknesses of both opposed arguments in a dialogue. Exercising the critical function means suspending the supportive and empathetic functions temporarily. But it especially requires the ability to suspend the supportive function. It involves a bracketing of your partisan feelings that push toward supporting your own point of view.

Critical doubt requires a neutral point of view in argument, neither positive nor negative. According to van Eemeren and Grootendorst (1984, 79), there are three possible attitudes a language user can have in a critical discussion—the positive point of view, the negative point of view, and the zero, or neutral, point of view. With respect to a given proposition at issue, a language user can be committed to it (positive point of view), committed to the opposite of it (negative point of view), or not committed one way or the other with respect to it (neutral point of view). The exercise of critical doubt requires at least a temporary adoption or presumption of the neutral point of view. But bias is precisely the thing that interferes with or prevents this suspension of one's positive or negative point of view.

Good bias is operative when you are defending your point of view appropriately. Bad bias is operative when this pro-argumentation attitude is not appropriate for the task in hand and gets in the way. This can happen in two ways. The supportive function can get in the way of the empathic function, when it interferes with your ability to put yourself inside your opponent's point of view. This type of bias is called *empathetic interference bias*. But the more usual type of bias, called *critical interference bias*, occurs when the supportive function interferes with the critical function. This occurs where you should adopt a neutral attitude of critical doubt, exploring the stronger or weaker points in your own or your opponent's argument, but the supportive function is so strong that it blocks this ability.

To sum up, there are two types of bad bias. One, the weaker bias, comes about through mixing functions in a dialogue. The other, the stronger bias, comes about through an illicit shift to another type of dialogue. In the first kind of bias, the problem is that an arguer gets carried away with the supportive function. When this becomes extreme, it may involve a shift, for example, from a critical discussion to a quarrel. In such a case, the weaker bias has hardened into the stronger bias. In the next section, various degrees of this hardening of bias will be discussed.

Being biased in argument is not a fallacy. But bad bias is connected with many fallacies—for example, the *ad hominem* fallacy. In case 6.7 Wilma was attacked by an *ad hominem* argument because she was supposed to be engaged in a dialogue on the issue of acid rain, a dialogue that was to look at the arguments on both sides of the issue in an open manner, whereas her unannounced connection with the coal industry raised questions about her sincerity and openness to both sides in the dialogue. There was good evidence of bad bias on Wilma's part in this case, and therefore the *ad hominem* argument raised against her was a reasonable and appropriate criticism. The *ad hominem* argument, then, functions as an allegation of bias that can be reasonable as an argument in some cases and fallacious in other cases.

The more typical charge of internal bias tends to be of critical interference bias. Many emotional fallacies are based on this type of bias—the problem is that the supportive function is often accompanied by strong emotions and "gut feelings" on an issue that make it difficult to bracket one's positive or negative attitude temporarily in order to adopt the neutral attitude necessary for critical doubt.

Empathic interference bias tends to be associated with the straw man fallacy and with problems associated with the fair evaluation or identification of unexpressed premises in argumentation. If your own implicit commitments are deep and emotional and you are strongly opposed to your opponent's point of view, it becomes more difficult to interpret or make presumptions about that opponent's commitments in a fair, charitable, and constructive manner. This type of bias is also connected to the *ad hominem* argument in some cases. For example, in extreme cases of poisoning the well, one's opponents may be labeled "liars" or "devils" that can never be trusted.[4] A total failure of empathy of this type makes a critical discussion impossible.

The Group Quarrel

In the simplest case, the quarrel is a personal conflict between two individuals, but this type of quarrel often has a way of becoming systematized and institutionalized, once set in the framework of a larger group conflict. When this happens, each group may either have or develop its own "official" position on the issue, and on other related issues, and even develop its own argumentative vocabulary for describing the conflict. In many cases, the group is defined by common interests or characteristics of its members—they could be national groups, professional groups, or even groups who have banded together because of some common interest, grievance, or cause. When a quarrel is set in this type of group framework, it may be called a *group quarrel*. Group quarrels have an institutional nature and tend to be much more stable over time than individual, personal quarrels.

An example would be a border dispute between two countries. As each side begins to perceive the other side as "the enemy," each develops its own one-sided terminology to describe the conflict and the opposing group. Each side describes its own soldiers as "freedom fighters" and the opposed side as "terrorists." Popular, derogatory terms like "gooks" or "huns" may commonly be used to describe the individuals on the other side, as though they were either not human or, if human, brutal and nasty.

The organized group quarrel is built around a rhetoric of group belonging and an emotional opposition to another group with whom the first group is, or is alleged to be, in conflict. Grievances against the "hostile" group are ritualized, and this group is portrayed as inherently untrustworthy, dishonest, and unreasonable. Whatever the other side says, their arguments are reflexively interpreted by the group as attempted deceptions or propaganda. The organized group quarrel is equivalent as a type of dialogue to *adversary argument*, defined by Flowers, McGuire, and Birnbaum (1982) as

---

4. On the use of the "lying devils" terminology to classify one's opponents in dialogue see Walton 1989 (153).

a dialogue where both parties intend to remain adversaries, neither party is really willing to be persuaded rationally, and both parties have the goal of making their own side look good and the opposing side look bad. The *modus operandi* of each party in this type of argument exchange is the *ad hominem* attack—the one side attacks the other for having committed "bad" actions, and the other side "counterblames" with *tu quoque ad hominem* attacks of the same sort.

Flowers, McGuire, and Birnbaum (1982, 283) also note that in this type of adversarial dialogue, terminology is used in an institutionalized way that always supports one's own side and carries negative implications for the position of the other side. This terminology of "good guys" and "bad guys" seems to be a requirement of group solidarity and identification in the institutionalized group quarrel.

The organized group quarrel requires that every member of the group adopt a particular point of view or position, dividing the world into those who accept this point of view and belong to the group, and outsiders or "enemies." In other words, the group loyalty forbids a neutral position. Those not in the group are defined as "enemies" or outsiders, potential threats. Thus, the group member always has a bias, but it is not seen by the group as a bad bias, only as reflecting the partisan nature of defending the group point of view.

Propaganda is a technique used by one side in an organized group quarrel to reaffirm group values and solidarity and to respond to influences outside the group. Propaganda is a negative kind of dialogue in that it excludes critical discussion, but it is not an inherently bad or immoral kind of argumentation *per se*. In national emergencies, or for other practical reasons—for example, danger to human life—it may be necessary for an organized group to engage in propaganda. Although critical discussion is an important part of open political debate in a democratic country, national interests can conflict with open political debate on an issue in some cases.

On the other hand, it is a requirement of a real democracy that significant elements of critical discussion be involved in serious political debating. A speaker in a political discussion should be willing to change his position if offered convincing arguments by the other side. If this does not appear to be the case in a political debate, the audience will judge that the speaker is merely engaging in eristic propaganda instead of genuine critical discussion. But political argumentation in election speeches and the like is often hard to classify exactly, because all political debate is a mixed type of dialogue. It is for exactly this reason that people are generally mistrustful of political argumentation and inclined to presume that it should usually be classified as propaganda or eristic dialogue.

Propaganda is a technique of argumentation in an eristic dialogue context used to influence a group of people to act in a particular way in accordance with the position or perceived requirements of the organized group. Propaganda causes the group to act, or prepares the way for action, by arousing their emotions in a quarreling type of dialogue. Most often exploited as arguments are the *argumentum ad populum*, which appeals to group identification and solidarity, and the *argumentum ad hominem*, which attacks the opposition as hostile, blameworthy, and dishonest opponents who cannot be reached by reasoning. Critical discussion is not a useful type of dialogue for the purposes of propaganda and is even at odds with it in most cases.

Propaganda works best when it is a double dialogue. On the surface there is a pretense of a critical discussion, but under the surface the dialogue is eristic argumentation that always promotes the point of view of one's own side in a partisan/biased manner. The aim is to defeat the opposed point of view. But if propaganda is too obviously biased and one-sided, it will not function effectively. Propaganda is not inherently immoral, but it is based on an underlying duplicity of types of dialogue—it is a type of argumentation that is designed to be deceptive. To be successful, it must not too easily appear to be what it really is.

Recognizing propaganda in particular cases requires a sensitivity to dialectical shifts and an ability to articulate the real position of a speaker. When propaganda is used to support the interests of groups we oppose or recognize as bad, it is clearly evident. But when propaganda is used to support our own group, or positions we support and approve of, it is much more difficult to detect. It is often harder to recognize your own bias than someone else's, especially where it is a group bias and your own interests are involved.

When one is caught up in a quarrel, it becomes difficult to disengage sufficiently from those emotions associated with group solidarity and personal attack to permit the attitude of critical doubt necessary for critical discussion.

What seems to be happening currently is that group quarrels are being instigated as a way of gaining political concessions to advance the interests of members of the group. An emotion of anger is simulated, along with other heavily emotional appeals, supposedly in response to bad actions, transgressions, or repressions carried out against the group. Thus, the group claim is to collective victimhood at the hands of the larger society, which must now make financial reparations for its collective guilt.

Such a group may not be a real community. They may be tied together, for example, only by virtue of some biological characteristic they share. Yet by arousing emotions of guilt and blame, they advance *ad hominem* attacks against their perceived enemies and make *ad baculum* threats as demands for concessions. One of their key tactics is the *argumentum ad misericordiam*, used to portray themselves as pathetic victims in need of reparations and special treatment. And the *argumentum ad populum* is used to affirm an emotional group solidarity and to divide the world into the group and "outsiders" or enemies who are supposed to be constantly ready to commit injustices against the group.

Once this kind of group quarrel is instigated and begins to take hold, a dogmatic attitude can set in. The group develops a systematic vocabulary and a "worldview," portraying itself as good and its perceived enemies as bad. History is reinterpreted as a story of the "oppression" and "liberation" of the group. The final step is fanaticism that leaves no room for questioning the group dogmas.

Dogmatism, Prejudice, and Fanaticism

Dogmatism, prejudice, and fanaticism are extreme, hardened forms of bad bias. Bias is *hardened* when the relationships among the three functions of argument in dialogue (supportive function, empathetic function, and critical function) are impaired and the proper balance required for good dialogue is destroyed.

Dogmatism is the refusal to go outside of the shell of the supportive function and deal with criticisms or counterarguments properly. A dogmatic attitude is similar to a quarrelsome attitude except that it is more defensive than aggressive. The dogmatic arguer will never alter any of the commitments in his fixed position, even when presented with convincing arguments. He may make minor alterations around the edges of his central position, but these will appear to him to be defensive in nature. Dogmatism is a hardened form of critical interference bias.

Prejudice is the failure to look at the other side of an issue or an inability to appreciate an opposing position. The prejudiced person does not really hear out the arguments for the other side and so does not enter into real dialogue with his opponent. He has made up his mind in advance and so never really understands the position of the other side. Prejudice is a hardened form of empathic interference bias.

Fanaticism is extremely hardened dogmatism where a systematic position or "world view" has been fixed into place so firmly that no room for any argument is left.[5] Fanaticism is an attitude that may be identified with the group quarrel.

You can argue with a dogmatic or prejudiced individual, but it will be tough going. But there is no point in trying to argue with a fanatic, for such a person does not perceive himself as arguing at all. He merely tries to "convert" you, or to use propagandistic arguments he is familiar with or trained to use. He is not really arguing in a critical discussion.

Dogmatism and prejudice are hard to crack, but fanaticism is more difficult still. A critical discussion requires openness to opposing arguments and points of view. Dogmatism and prejudice are forms of partial closure. Fanaticism is complete closure. You may think you are "reasoning" with a fanatic, but for him it is not a critical discussion at all—he sees it as a group quarrel when he engages in dialogue with anyone who is outside his group. Nevertheless, for strategic reasons, such a person will try to appear to be engaging in critical discussion.

In a critical discussion, you are always partial and have a positive bias, because you are trying to persuade the other party that your point of view is right. This is the supportive function of pro-argumentation. But sometimes you have to modify your position, if presented with good arguments that go against it. Dogmatic or prejudiced arguers won't do this, or have difficulty doing it.

Also, sometimes you have to suspend your positive attitude and try to adopt a neutral attitude of critical doubt. Bias tends to interfere with this suspension or make it more difficult. This interference is associated with dogmatism, prejudice, and fallacies.

Two requirements of successfully engaging in a critical discussion are the attitudes of openness and balance, both of which involve a readiness to look at the other person's arguments and take them into account. You have to push for your own position but also be sensitive to the weaknesses in it and to the strengths of the arguments on the opposed side. Balance is the fine tuning of these supportive, critical, and empathetic functions.

You also have to be open and flexible, to have a certain amount of

5. See Jay Newman, Fanatics and Hypocrites (Buffalo: Prometheus Books, 1986).

"give," or judgment of when to concede and when to resist. Dogmatic and prejudiced arguers have difficulties with balance and openness. Fanatics have no balance and no openness—they start out with a position like a concrete fortification. They can't really enter into a critical discussion at all, and never have any intention of doing so.

The poisoning-the-well *ad hominem* argument makes the allegation that the person criticized is dogmatic or prejudiced, implying that he can never be trusted to enter into a sincere critical discussion of an issue. The circumstantial type of *ad hominem* argument makes the allegation that the person criticized is hypocritical, implying that he cannot be trusted as a sincere advocate of the argument he expounds.[6] A hypocritical arguer is an arguer who expressly puts forward a proposition as something he advocates, but his personal circumstances show that he is in fact not committed to this proposition personally at all (Walton 1985). For example, a hypocritical arguer might verbally expound a principle that says, "Everyone ought to refrain from smoking," while he himself smokes—see case 6.4 for an elaboration of this type of *ad hominem* argumentation.

Hypocrisy is not a type of bias. It is an allegation of insincerity in entering into an argument, based on an alleged inconsistency in the arguer's commitments. One could say that the dogmatic or prejudiced arguer is too inflexible, while the hypocritical arguer is too flexible; one is too rigid and the other is not rigid enough. Thus the circumstantial *ad hominem* and the poisoning-the-well type of *ad hominem* are two distinctively different subspecies of personal attack in argument.

The poisoning-the-well variant of the *ad hominem* fallacy often occurs where a legitimate question of bias is exaggerated as a much stronger allegation of prejudice, dogmatism, or bias. For example, in a recent case there was a legitimate question of whether a judge being considered for the Supreme Court had a political bias that might interfere with his legal reasoning. However, his critics phrased the accusation in much stronger terms, calling him a "right-wing zealot" whose political bias would "dictate" his legal opinions. The problem with many *ad hominem* arguments is that they have a reasonable basis as weak presumptive arguments that raise critical questions, but they become fallacious when the attack is pressed ahead too hard. Such weak presumption-based arguments often cannot bear up under this kind of weight. A critical questioning of an arguer's bias

6. Ibid.

is one thing, but calling the arguer a "fanatic" or "zealot" is a much stronger and more serious allegation.

Fanaticism is an attitude that is essentially associated with the group quarrel. The fanatic sees everyone who does not belong to the group as an outsider, who is either an enemy of the group or needs to be "converted" into a member of the group. Thus, for the fanatic, there is no place for critical discussion. The way to relate verbally to anyone outside the group is through the group propaganda, which takes a quarrelsome stance towards all outsiders. Thus, fanaticism is always based on some official or established group position, a systematic "world view."

Dogmatism is often associated with some systematic group position, for example, in religious dogma. A person may be said to be dogmatic who is slow to adapt to new ideas and who clings too stubbornly to his established commitments even in the face of contrary evidence. But a dogmatic attitude is not necessarily tied to the group quarrel dialogue. For the fanatic, every verbal exchange with anyone outside the group must be an essentially quarrelsome dialogue—a kind of propaganda that defends or preserves the values and interests of the group against its enemies.

BIBLIOGRAPHY

Ackermann, Alfred S. E. 1970. *Popular fallacies*. 4th ed. Detroit: Gale Research Co.

Anscombe, G.E.M. 1957. *Intention*. Oxford: Blackwell.

Aristotle. 1928. *The works of Aristotle translated into English*. Ed. W. D. Ross. Oxford: Oxford University Press.

———. 1955. *On sophistical refutations*. Trans. E. S. Forster. Cambridge: Harvard University Press, Loeb Classical Library Edition; London: William Heinemann.

Bach, Kent, and Robert M. Harnish. 1979. *Linguistic communication and speech acts*. Cambridge: MIT Press.

Bailey, F. G. 1983. *The tactical uses of passion*. Ithaca: Cornell University Press.

Barth, E. M., and E. C. W. Krabbe. 1982. *From axiom to dialogue*. New York: De Gruyter.

Barth, E. M., and J. L. Martens. 1977. Argumentum ad hominem: From chaos to formal dialectic. *Logique et Analyse* 77–78: 76–96.

Blair, J. Anthony. 1988. What is bias? In *Selected issues in logic and communication*, ed. Trudy Govier, 93–103. Belmont, Calif.: Wadsworth.

Bochenski, J. M. 1974. An analysis of authority. In *Authority*, ed. Frederick J. Adelman, 56–85. The Hague: Martinus Nijhoff.

Brinton, Alan. 1985. A rhetorical view of the *ad hominem*. *Australian Journal of Philosophy* 63:50–63.

———. 1986. Ethotic argument. *History of Philosophy Quarterly* 3:245–57.

———. 1987. Ethotic argument: Some uses. In *Argumentation: Perspectives and approaches*, ed. Frans H. van Eemeren, Rob Grootendorst, J. Anthony Blair, and Charles A. Willard, 246–54. Dordrecht: Foris Publications.

———. 1988a. Pathos and the 'appeal to emotion': An Aristotelian analysis. *History of Philosophy Quarterly* 5:207–19.

———. 1988b. Appeal to the angry emotions. *Informal Logic* 10:77–78.

Callahan, Sidney. 1988. The role of emotion in ethical decisionmaking. *Hastings Center Report* 18:9–14.

Canada: House of Commons debates (Hansard). Ottawa: The Queen's Printer.

Carlson, Lauri. 1983. *Dialogue games*. Dordrecht: Reidel.

Castell, Alburey. 1935. *A college logic*. New York: Macmillan.

Cederblom, Jerry, and David W. Paulsen. 1982. *Critical reasoning*. Belmont, Calif.: Wadsworth.

Clark, Erik. 1988. *The want makers*. London: Hodder & Stoughton.

Copi, Irving M. 1986. *Introduction to logic*. 7th ed. New York: Macmillan [6th ed., 1982].

Copi, Irving M., and Carl Cohen. 1990. *Introduction to logic*. 8th ed. New York: Macmillan.

Cuomo, Mario L. 1984. Religious belief and public morality. *The New York Review of Books* 31 (25 October): 32–37.

Damer, Edward T. 1980. *Attacking faulty reasoning*. Belmont, Calif.: Wadsworth.

Degnan, Ronan E. 1973. Evidence. In *Encyclopaedia Britannica*, 15th ed., vol. 8, 905–16.

De George, Richard T. 1985. *The nature and limits of authority*. Lawrence: University Press of Kansas.

DeMorgan, Augustus. 1847. *Formal logic*. London: Taylor and Walton.

Donohue, William A. 1981a. Analyzing negotiation tactics: Development of a negotiation interact system. *Human Communication Research* 7:237–87.

———. 1981b. Development of a model of rule use in negotiation interaction. *Communication Monographs* 48: 106–20.

Edwards, Paul. 1967. Common consent arguments for the existence of God. In *Encyclopedia of philosophy*, ed. Paul Edwards, vol. 2. New York: Macmillan.

Engel, Morris S. 1976. *With good reason: An introduction to informal fallacies*. New York: St. Martin's Press.

Evans, J.D.G. 1977. *Aristotle's concept of dialectic*. London: Cambridge University Press.

Fearnside, W. Ward, and William B. Holther. 1959. *Fallacy: The counterfeit of argument*. Englewood Cliffs, N.J.: Prentice-Hall.

Fineman, Howard. 1988. Poppy the populist: A onetime preppy talks to the folks. *Newsweek*, 7 November, 58–60.

Fischer, David Hackett. 1970. *Historians' fallacies*. New York: Harper & Row.

Fisher, Roger, and William Ury. 1983. *Getting to yes: Negotiated agreement without giving in*. London: Hutchinson.

Flowers, Margot, Rod McGuire, and Lawrence Birnbaum. 1982. Adversary arguments and the logic of personal attacks. In *Strategies for natural language processing*, ed. Wendy G. Lehnert and Martin H. Ringle, 275–94. Hillsdale, N.J.: Lawrence Erlbaum Associates.

Freeman, James B. 1988. *Thinking logically: Basic concepts of reasoning.* Englewood Cliffs, N.J.: Prentice-Hall.

Gelman, David, and Mark Miller. 1987. A new attack on abortion. *Newsweek,* 2 February, 32.

Glucksberg, Sam, and Michael McCloskey. 1981. Decisions about ignorance: Knowing that you don't know. *Journal of Experimental Psychology: Human Learning and Memory* 7: 311–25.

Govier, Trudy. 1983. Ad hominem: Revising the textbooks. *Teaching Philosophy* 6:13–24.

———. 1985. *A practical study of argument.* Belmont, Calif.: Wadsworth.

———. 1987. *Problems in argument analysis and evaluation.* Dordrecht: Foris Publications.

Graham, Michael H. 1977. Impeaching the professional expert witness by a showing of financial interest. *Indiana Law Journal* 53:35–53.

Grice, Paul H. 1975. Logic and conversation. In *The Logic of Grammar,* ed. Donald Davidson and Gilbert Harman, 64–75. Encino, Calif.: Dickenson.

Groarke, Leo, and Christopher Tindale. 1986. Critical thinking: How to teach good reasoning. *Teaching Philosophy* 9:301–18.

Hamblin, Charles L. 1970. *Fallacies.* London: Methuen. [Reprint. 1986. Newport News, Va.: Vale Press.]

———. 1971. Mathematical models of dialogue. *Theoria* 37:130–55.

Harrah, David. 1971. Formal message theory. In *Pragmatics of natural languages,* ed. Yehoshua Bar-Hillel. Dordrecht: Reidel.

———. 1976. Formal message theory and non-formal discourse. In *Pragmatics of language and literature,* ed. Teun A. van Dijk. Amsterdam: North-Holland.

Hart, H.L.A., and A.M. Honore. 1969. *Causation in the law.* London: Oxford University Press.

Hoffman, Herbert C. 1979. The cross-examination of expert witnesses. In *Planning, Zoning, and Eminent Domain Institute,* 313–49.

Ilbert, Sir Courtney. 1960. Evidence. In *Encyclopaedia Britannica,* 11th ed., vol. 10, 11–21.

Imwinkelried, Edward J. 1981. *Scientific and expert evidence.* New York: Practicing Law Institute.

———. 1986. Science takes the stand: The growing misuse of expert testimony. *The Sciences* 26:20–25.

Jacquette, Dale. 1989. The hidden logic of slippery slope arguments. *Philosophy and Rhetoric* 22:59–70.

Johnson, Ralph H., and J. Anthony Blair. 1983. *Logical self-defense.* 2d ed. Toronto: McGraw-Hill Ryerson.

Johnstone, Henry W., Jr. 1970. Philosophy and *argumentum ad hominem* revisited. *Revue Internationale de Philosophie* 24:107–16.

———. 1978. *Validity and rhetoric in philosophical argument.* University Park, Pa.: Dialogue Press of Man and World.

Kahane, Howard. 1982. *Logic and philosophy.* 4th ed. Belmont, Calif.: Wadsworth.

Kapp, Ernst. 1942. *Greek foundations of traditional logic.* New York: Columbia University Press.

Kennedy, John F. 1961. For the freedom of man we must all work together. *Vital Speeches of the Day* 27 (1 February): 226–27.

Kenny, Anthony. 1983. The expert in court. *Law Quarterly Review* 99:197–216.

Kerferd, G. B. 1981. *The Sophistic movement.* Cambridge: Cambridge University Press.

Kielkopf, Charles. 1980. Relevant appeals to force, pity, and popular pieties. *Informal Logic* 2:2–5.

Krabbe, Erik C. W. 1985. Formal systems of dialogue rules. *Synthese* 63:295–328.

Kreckel, Marga. 1981. *Communicative acts and shared knowledge in natural discourse.* London: Academic Press.

Lewis, Charlton T., and Charles Short. 1969. *A Latin dictionary.* Oxford: Clarendon Press.

Locke, John. [1690] 1961. *An essay concerning human understanding.* Ed. John W. Yolton. 2 vols. London: Dent.

Loftus, Elizabeth F. 1979. *Eyewitness testimony.* Cambridge: Harvard University Press.

Lorenzen, Paul. 1969. *Normative logic and ethics.* Mannheim: Bibliographisches Institut.

Mackenzie, J. D. 1981. The dialectics of logic. *Logique et Analyse* 94:159–77.

Manor, Ruth. 1981. Dialogues and the logics of questions and answers. *Linguistische Berichte* 73:1–28.

———. 1982. Pragmatics and the logic of questions and answers. *Philosophica* 29:45–96.

Massey, Gerald. 1981. The fallacy behind fallacies. *Midwest Studies in Philosophy* 6:489–500.

Michalos, Alex C. 1970. *Improving your reasoning.* Englewood Cliffs, N.J.: Prentice-Hall.

Moore, Christopher W. 1986. *The mediation process.* San Francisco: Jossey-Bass Publishers.

Moore, James A., James A. Levin, and William C. Mann. 1977. A goal-oriented model of human dialogue. *American Journal of Computational Linguistics* 67:1–54.

Perelman, Chaim. 1982. *The realm of rhetoric.* Notre Dame: University of Notre Dame Press.

Perelman, Chaim, and L. Olbrechts-Tyteca. 1969. *The new rhetoric: A treatise on argumentation.* Trans. John Wilkinson and Purcell Weaver. Notre Dame: University of Notre Dame Press.

Reiter, Raymond. 1987. Nonmonotonic reasoning. *Annual Review of Computer Science* 2:147–86.

Rescher, Nicholas. 1976. *Plausible reasoning.* Assen-Amsterdam: Van Gorcum.

———. 1977. *Dialectics.* Albany: State University of New York Press.

———. 1988. *Rationality.* Oxford: Oxford University Press.

Robinson, Richard. 1953. *Plato's earlier dialectic.* Oxford: Oxford University Press.

———. 1971. Arguing from ignorance. *Philosophical Quarterly* 21:97–108.

Samuelson, Robert J. 1988. The elderly aren't needy. *Newsweek,* 21 March, 68.

Saunders, John B., ed. 1970. *Words and phrases legally defined.* Vol. 5. London: Butterworths.

Searle, John. 1969. *Speech acts.* Cambridge: Cambridge University Press.

———. 1975. Indirect speech acts. In *Syntax and semantics,* ed. Peter Cole and Jerry L. Morgan, vol. 3, 59–82. New York: Academic Press.

Shepherd, Robert Gordon, and Erich Goode. 1977. Scientists in the popular press. *New Scientist* 76:482–84.

Sidgwick, Alfred. 1914. *Elementary logic.* Cambridge: Cambridge University Press.

Solomon, Robert C. 1988. On emotions as judgments. *American Philosophical Quarterly* 25:183–91.

Thouless, Robert H. 1930. *Straight and crooked thinking.* London: English Universities Press.

Ullman-Margalit, Edna. 1983. On presumption. *The Journal of Philosophy* 80:143–63.

Van de Vate, Dwight, Jr. 1975a. The appeal to force. *Philosophy and Rhetoric* 8:43–60.
———. 1975b. Reasoning and threatening: A reply to Yoos. *Philosophy and Rhetoric* 8:177–79.
van Eemeren, Frans H., and Rob Grootendorst. 1984. *Speech acts in argumentative discussions*. Dordrecht: Foris Publications.
———. 1987. Fallacies in pragma-dialectical perspective. *Argumentation* 1:283–301.
———. 1990. Indirect speech acts in a discourse text.
Van Eemeren, Frans H., and Tjark Kruiger. 1987. Indentifying argumentation schemes. In *Argumentation: Perspectives and approaches*, ed. Frans H. Van Eemeren, Rob Grootendorst, J. Anthony Blair, and Charles Willard. Dordrecht: Foris Publications.
Waller, Bruce N. 1988. *Critical Thinking*. Englewood Cliffs: Prentice-Hall.
Walton, Douglas N. 1980. Why is the *ad populum* a fallacy? *Philosophy and Rhetoric* 13:264–78.
———. 1982. *Topical relevance in argumentation*. Amsterdam: John Benjamins Publishing Co.
———. 1984. *Logical dialogue-games and fallacies*. Lanham, Md.: University Press of America.
———. 1985. *Arguer's position*. Westport, Conn.: Greenwood Press.
———. 1987a. The *ad hominem* argument as an informal fallacy. *Argumentation* 1:317–32.
———. 1987b. *Informal fallacies*. Amsterdam: John Benjamins Publishing Co.
———. 1988. Burden of proof. *Argumentation* 2:233–54.
———. 1989a. *Informal logic*. Cambridge: Cambridge University Press.
———. 1989b. *Question-reply argumentation*. New York: Greenwood Press.
———. 1990. *Practical reasoning*. Savage, Md.: Rowman and Littlefield.
Walton, Douglas N., and Lynn M. Batten. 1984. Games, graphs, and circular arguments. *Logique et Analyse* 106: 133–64.
Walton, Douglas N., and Erik C. W. Krabbe. 1990. *Commitment in dialogue*.
Waterman, Donald A. 1986. *A guide to expert systems*. Reading, Mass.: Addison-Wesley.
Weber, O. J. 1981. Attacking the expert witness. *Federation of Insurance Counsel Quarterly* 31:299–313.
Weddle, Perry. 1978. *Argument: A guide to critical thinking*. New York: McGraw-Hill.
Whately, Richard. 1836. *Elements of logic*. New York: William Jackson.
———. [1846] 1963. *Elements of rhetoric*. Ed. Douglas Ehninger. Carbondale: Southern Illinois University Press.
Wilensky, R. 1983. *Planning and understanding: A computational approach to human reasoning*. Reading, Mass.: Addison-Wesley.
Wilson, Patrick. 1983. *Second-hand knowledge: An inquiry into cognitive authority*. Westport, Conn.: Greenwood Press.
Wilson, Thomas. 1552. *The rule of reason*. London: Grafton.
Windes, Russel R., and Arthur Hastings. 1965. *Argumentation and advocacy*. New York: Random House.
Woods, John. 1987. Ad baculum, self-interest, and Pascal's wager. In *Argumentation: Across the lines of discipline*, ed. Frans H. van Eemeren, Rob Grootendorst, J. Anthony Blair, and Charles A. Willard, 343–49. Dordrecht: Foris Publications.
Woods, John, and Douglas N. Walton. 1974. Argumentum ad verecundiam. *Philosophy and Rhetoric* 7:135–53.
———. 1976. Ad baculum. *Grazer Philosophische Studien* 2:133–40.

————. 1977. Ad hominem. The Philosophical Forum 8:1–20.

————. 1978. The fallacy of ad ignorantiam. Dialectia 32:87–99.

————. 1982. Argument: The logic of the fallacies. Toronto: McGraw-Hill Ryerson.

————. 1989. Fallacies: Selected papers, 1972–1982. Dordrecht: Foris Publications.

Wreen, Michael J. 1987. Yes, Virginia, there is a Santa Claus. Informal Logic 9:31–39.

————. 1988a. Admit no force but argument. Informal Logic 10:89–96.

————. 1988b. May the force be with you. Argumentation 2:425–40.

————. 1989. A bolt of fear. Philosophy and Rhetoric 22:131–40.

Yoos, George E. 1975. A critique of Van de Vate's "The appeal to force." Philosophy and
    Rhetoric 8:172–76.

Younger, Irving. 1982. A practical approach to the use of expert testimony. Cleveland
    State Law Review 31:1–42.

INDEX

initial, 97
local weight of, 35
masked lack of evidence, 129
permissible, 63
pragmatic logic of, 60
reasonable, 63
reverses the burden of proof, 60
six noteworthy bases for making, 59
tentative, 95
three types of, 63
vs. presupposition, 56–57
weak, 62
weight of, 38, 222
  evaluation of, 197
  shifting, 95, 115, 253
Prior Analytics, 71
propaganda, 274–75
propositions, 12, 36
  guides to action, 106
  internal stock, 12
  unproven, 5

quarrel, 236, 257, 266, 275
  adversarial, 214
  attitude in, 267
  cathartic function of, 21, 215
  compared to debate, 216
  contrasted to critical discussion, 21
  defined, 21
  deterioration into, 192, 208, 210, 218
  group, 215, 237, 240, 246–47, 250, 260,
    264, 275
    defined, 273
    organized, 273–74
  personal, 3, 214
  shift to, 264
  valuable benefits of, 22
question
  complex, 38
  loaded, 38, 132, 135
  open, 253
question-reply interaction, 13, 35

reasoning
  backwards, 12
  concept of, 11–13
  deductive, 14–15
  defeasible, 43, 198
  interactive, 11, 13–15, 33
    dialectical, 14

evaluation of, 15
  two person, 15
  knowledge-based, 11, 14–15, 32, 72–73
  mololectical, 14
  practical, 68, 101, 113
    based on, 90
    defined, 19
    goal directed, 110, 168
    guided by, 256
  presumptive, 27, 32, 34
    defeasible, 18–19, 96
    negative logic of, 61
    related to hard evidence, 62
  scientific, 72
  source-based, 198
refutation, weak, 207
relevance, 29
  failure of, 74
  normative, 75
  pragma-dialectical concept of, 74
Reuther, Walter, 100–101, 250
Rhetoric, 67, 91
Roberts, Oral, 184–85
Rodriguez, John R., 110–11
rule, 36
  commitment, 72
  for successful communication, 187
  legal . . . of evidence, 195
  locution, 72
  of critical discussion, 35, 151
  of debate, 112
  of dialogue, 72, 260, 263–64
  of presumptive reasoning, 35
  opinion, 46
  plea of exemption to, 116–18
  procedural . . . of dialogue, 33, 65, 71,
    73, 215
    for a critical discussion, 74
    violation of . . . of critical discussion,
      16
    violation of . . . of dialogue, 18, 161
    of negotiation, 151
  win-loss, 20, 72

Samuelson, Robert J., 136–37
Saunders, John B., 162
scandalmonger, 205
scaremongering, 174–76, 242
Searle, John R., 169, 173
Seneca, 2
sentiments, 106, 108
Shepherd, Robert Gordon, 52